SCHAUM'S *Easy* OUTLINES

DISCRETE
MATHEMATICS

Other Books in Schaum's Easy Outlines Series Include:

SCHAUM'S *Easy* OUTLINES

DISCRETE

MATHEMATICS

BASED ON SCHAUM'S
Outline of Theory and Problems of Discrete
Mathematics, Second Edition
BY SEYMOUR LIPSCHUTZ, Ph.D.
AND MARC LARS LIPSON, Ph.D.

ABRIDGEMENT EDITOR:
GEORGE J. HADEMENOS, Ph.D.

SCHAUM'S OUTLINE SERIES

McGRAW-HILL

New York   Chicago   San Francisco   Lisbon   London   Madrid
Mexico City   Milan   New Delhi   San Juan
Seoul   Singapore   Sydney   Toronto

SEYMOUR LIPSCHUTZ is on the faculty of Temple University and formerly taught at the Polytechnic Institute of Brooklyn. He received his Ph.D. in 1960 at Courant Institute of Mathematical Sciences of New York University. He is one of Schaum's most prolific authors and has written, among others, *Schaum's Outlines of Beginning Linear Algebra, Probability, Discrete Mathematics, Set Theory, Finite Mathematics*, and *General Topology*.

MARC LARS LIPSON teaches at the University of Georgia. He received his Ph.D. in finance in 1994 from the University of Michigan, and he is the co-author with Seymour Lipschutz of *Schaum's Outlines of Discrete Mathematics and Probability*.

GEORGE J. HADEMENOS has taught at the University of Dallas and done research at the University of Massachusetts Medical Center and the University of California at Los Angeles. He holds a B.S. degree from Angelo State University and both M.S. and Ph.D. degrees from the University of Texas at Dallas. He is the author of several books in the *Schaum's Outline* and *Schaum's Easy Outline* series.

1 2 3 4 5 6 7 8 9 DOC DOC 0 9 8 7 6 5 4 3 2

ISBN 0-07-139877-5

Library of Congress Cataloging-in-Publication Data applied for.

Sponsoring Editor: Barbara Gilson
Production Supervisors: Tama Harris and Clara Stanley
Editing Supervisor: Maureen B. Walker

McGraw-Hill

A Division of The McGraw-Hill Companies

Contents

SCHAUM'S *Easy* OUTLINES

DISCRETE
MATHEMATICS

Chapter 1
SET THEORY

IN THIS CHAPTER:

- ✔ *Sets and Elements*
- ✔ *Universal Set and Empty Set*
- ✔ *Subsets*
- ✔ *Venn Diagrams*
- ✔ *Set Operations*
- ✔ *Algebra of Sets and Duality*
- ✔ *Finite Sets, Counting Principle*
- ✔ *Classes of Sets, Power Sets, Partitions*

Sets and Elements

A *set* may be viewed as a collection of objects, the *elements* or *members* of the set. We ordinarily use capital letters, $A, B, X, Y, \ldots$, to denote sets, and lowercase letters, $a, b, x, y, \ldots$, to denote elements of sets. The statement "p is an element of A," or equivalently, "p belongs to A," is written

$$p \in A$$

1

The statement that p is not an element of A, that is, the negation of $p \in A$, is written

$$p \notin A$$

The fact that a set is completely determined when its members are specified is formally stated as the principle of extension.

Principle of Extension:

Two sets A and B are equal if and only if they have the same members.

Specifying Sets

There are essentially two ways to specify a particular set. One way, if possible, is to list its members. For example,

$$A = \{a, e, i, o, u\}$$

denotes the set A whose elements are the letters a, e, i, o , u. Note that the elements are separated by commas and enclosed in braces { }. The second way is to state those properties which characterized the elements in the set. For example,

$$B = \{x: x \text{ is an even integer}, x > 0\}$$

reads "B is the set of x such that x is an even integer and x is greater than 0." It denotes the set B whose elements are the positive integers. A letter, in this case x, is used to denote a typical member of the set; the colon is read as "such that" and the comma as "and."

Solved Problem 1.1

    (*a*) The set A above can also be written as

        $A = \{x: x \text{ is a letter in the English alphabet}, x \text{ is a vowel}\}$

        Observe that b $\notin A$, e $\in A$, and p $\notin A$.

(b) We could not list all the elements of the above set B although frequently we specify the set by writing

$$B = \{2, 4, 6, \ldots\}$$

where we assume that everyone knows what we mean. Observe that $8 \in B$ but $-7 \notin B$.

(c) Let $E = \{x: x^2 - 3x + 2 = 0\}$. In other words, E consists of those numbers which are solutions of the equation $x^2 - 3x + 2 = 0$, sometimes called the *solution set* of the given equation. Since the solutions of the equation are 1 and 2, we could also write $E = \{1, 2\}$.

(d) Let $E = \{x: x^2 - 3x + 2 = 0\}$, $F = \{2, 1\}$ and $G = \{1, 2, 2, 1, 2\}$. Then $E = F = G$. Observe that a set does not depend on the way in which its elements are displayed. A set remains the same if its elements are repeated or rearranged.

You Need to Know

Some sets will occur very often in the text and so we use special symbols for them. Unless otherwise specified, we will let

N = the set of positive integers: 1, 2, 3,. . .
Z = the set of integers: . . . , –2, –1, 0, 1, 2,. . .
Q = the set of rational numbers
R = the set of real numbers
C = the set of complex numbers

Even if we can list the elements of a set, it may not be practical to do so. For example, we would not list the members of the set of people born in the world during the year 1976 although theoretically it is possible to compile such a list. That is, we describe a set by listing its elements only if the set contains a few elements; otherwise we describe a set by the property which characterizes its elements.

The fact that we can describe a set in terms of a property is formally stated as the *principle of abstraction*.

Principle of Abstraction:
Given any set U and any property P, there is a set A such that the elements of A are exactly those members of U which have the property P.

Universal Set and Empty Set

In any application of the theory of sets, the members of all sets under investigation usually belong to some fixed large set called the *universal set*. For example, in plane geometry, the universal set consists of all the points in the plane, and in human population studies, the universal set consists of all the people in the world. We will let the symbol,

$$U$$

denote the universal set unless otherwise stated or implied.

For a given set U and a property P, there may not be any elements of U which have property P. For example, the set

$$S = \{x : x \text{ is a positive integer}, x^2 = 3\}$$

has no elements since no positive integer has the required property.

The set with no elements is called the *empty set* or *null set* and is denoted by

$$\varnothing$$

There is only one empty set. That is, if S and T are both empty, then $S = T$ since they have exactly the same elements, namely, none.

Subsets

If every element in a set A is also an element of a set B, then A is called a *subset* of B. We also say that A is *contained* in B or that B *contains* A. This relationship is written

$$A \subseteq B \text{ or } B \supseteq A$$

Solved Problem 1.2 The set $E = \{2, 4, 6\}$ is a subset of the set $F = \{6, 2, 4\}$, since each number 2, 4, and 6 belonging to E also belongs to F. In fact, $E = F$. In a similar manner, it can be shown that every set is a subset of itself.

The following properties of sets should be noted:

(i) Every set A is a subset of the universal set U since, by definition, all the elements of A belong to U. Also, the empty set $\varnothing$ is a subset of A.

(ii) Every set A is a subset of itself since, trivially, the elements of A belong to A.

(iii) If every element of A belongs to a set B, and every element of B belongs to a set C, then clearly every element of A belongs to C. In other words, if $A \subseteq B$ and $B \subseteq C$, then $A \subseteq C$.

(iv) If $A \subseteq B$ and $B \subseteq A$, then A and B have the same elements, i.e., $A = B$. Conversely, if $A = B$, then $A \subseteq B$ and $B \subseteq A$ since every set is a subset of itself.

We state these results formally.

Theorem 1.1:

(i)   For any set A, we have $\varnothing \subseteq A \subseteq U$.
(ii) For any set A, we have $A \subseteq A$.
(iii) If $A \subseteq B$ and $B \subseteq C$, then $A \subseteq C$.
(iv) $A = B$ if and only if $A \subseteq B$ and $B \subseteq A$.

If $A \subseteq B$, then it is still possible that $A = B$. When $A \subseteq B$ but $A \neq B$, we say A is a *proper subset* of B. We will write $A \subset B$ when A is a proper subset of B. For example, suppose

$$A = \{1, 3\} \qquad B = \{1, 2, 3\} \qquad C = \{1, 3, 2\}$$

Then A and B are both subsets of C; but A is a proper subset of C, whereas B is not a proper subset of C since $B = C$.

Venn Diagrams

A Venn diagram is a pictorial representation of sets in which sets are represented by enclosed areas in the plane.

The universal set U is represented by the interior of a rectangle, and the other sets are represented by disks lying within the rectangle. If $A \subseteq B$, then the disk representing A will be entirely within the disk representing B as in Figure 1-1(a). If A and B are disjoint, i.e., if they have no elements in common, then the disk representing A will be separated from the disk representing B as in Figure 1-1(b).

However, if A and B are two arbitrary sets, it is possible that some objects are in A but not in B, some are in B but not in A, some are in both A and B, and some are in neither A nor B; hence in general we represent A and B as in Figure 1-1(c).

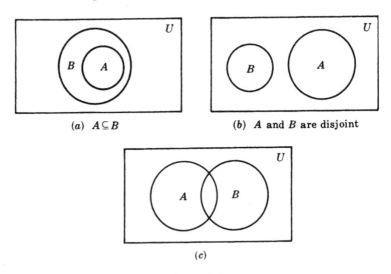

(a) $A \subseteq B$ (b) **A and B are disjoint**

(c)

Figure 1-1

Arguments and Venn Diagrams

Many verbal statements are essentially statements about sets and can therefore be described by Venn diagrams.

Hence Venn diagrams can sometimes be used to determine whether or not an argument is valid.

Solved Problem 1.3 Show that the following argument (adapted from a book on logic by Lewis Carroll, the author of *Alice in Wonderland*) is valid:

S_1: My saucepans are the only things I have that are made of tin.
S_2: I find all your presents very useful.
S_3: None of my saucepans is of the slightest use.

S: Your presents to me are not made of tin.

(The statements S_1, S_2, and S_3 denote the assumptions, and the statement S denotes the conclusion. The argument is valid if the conclusion S follows logically from the assumptions S_1, S_2, and S_3.)

By S_1, the tin objects are contained in the set of saucepans and by S_3, the set of saucepans and the set of useful things are disjoint: hence draw the Venn diagram of Figure 1-2.

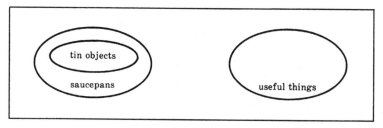

Figure 1-2

By S_2, the set of "your presents" is a subset of the set of useful things; hence draw Figure 1-3.

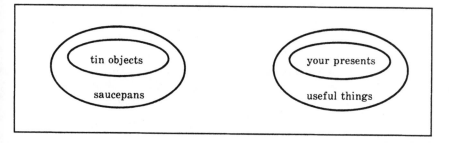

Figure 1-3

The conclusion is clearly valid by the above Venn diagram because the set of "your presents" is disjoint from the set of tin objects.

Set Operations

This section introduces a number of important operations on sets.

Union and Intersection

The *union* of two sets A and B, denoted by $A \cup B$, is the set of all elements which belong to A or to B; that is,

$$A \cup B = \{x: x \in A \text{ or } x \in B\}$$

Here "or" is used in the sense of and/or. Figure 1-4(*a*) is a Venn diagram in which $A \cup B$ is shaded.

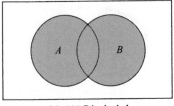
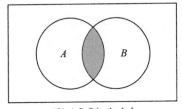

(*a*) $A \cup B$ is shaded   (*b*) $A \cap B$ is shaded

Figure 1-4

The *intersection* of two sets A and B, denoted by $A \cap B$, is the set of elements which belong to both A and B; that is,

$$A \cap B = \{x: x \in A \text{ and } x \in B\}$$

Figure 1-4(*b*) is a Venn diagram in which $A \cap B$ is shaded.

If $A \cap B = \varnothing$, that is, if A and B do not have any elements in common, then A and B are said to be *disjoint* or *nonintersecting*.

Solved Problem 1.4 Let $A = \{1, 2, 3, 4\}$, $B = \{3, 4, 5, 6, 7\}$, and $C = \{2, 3, 5, 7\}$. Find (*a*) $A \cup B$; (*b*) $A \cap B$; (*c*) $A \cup C$; and (*d*) $A \cap C$.

Solution.   (a) $A \cup B = \{1, 2, 3, 4, 5, 6, 7\}$
(b) $A \cap B = \{3, 4\}$
(c) $A \cup C = \{1, 2, 3, 4, 5, 7\}$
(d) $A \cap C = \{2, 3\}$

The operation of set inclusion is closely related to the operations of union and intersection, as shown by the following theorem.

Theorem 1.2: The following are equivalent: $A \subseteq B$, $A \cap B = A$, and $A \cup B = B$.

Complements

Recall that all sets under consideration at a particular time are subsets of a fixed universal set U. The *absolute complement* or simply *complement* of a set A, denoted by A^c, is the set of elements which belong to U but which do not belong to A; that is

$$A^c = \{x \colon x \in U, x \notin A\}$$

Figure 1-5(a) is a Venn diagram in which A^c is shaded.

The *relative complement* of a set B with respect to a set A or, simply, the *difference* of A and B, denoted by $A \backslash B$, is the set of elements which belong to A but which do not belong to B; that is

$$A \backslash B = \{x \colon x \in A, x \notin B\}$$

The set $A \backslash B$ is read "A minus B". Figure 1-5(b) is a Venn diagram in which $A \backslash B$ is shaded.

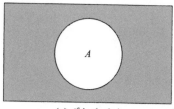

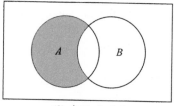

(a) A^c is shaded   (b) $A \backslash B$ is shaded

Figure 1-5

Symmetric Difference

The *symmetric difference* of sets A and B, denoted by $A \oplus B$, consists of those elements which belong to A or B but not to both; that is,

$$A \oplus B = (A \cup B) \setminus (A \cap B)$$

One can also show that

$$A \oplus B = (A \setminus B) \cup (B \setminus A)$$

For example, suppose $A = \{1, 2, 3, 4, 5, 6\}$ and $B = \{4, 5, 6, 7, 8, 9\}$. Then,

$$A \setminus B = \{1, 2, 3\}, \quad B \setminus A = \{7, 8, 9\} \text{ and so } \quad A \oplus B = \{1, 2, 3, 7, 8, 9\}$$

Figure 1-6 is a Venn diagram in which $A \oplus B$ is shaded.

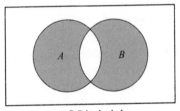

$A \oplus B$ is shaded

Figure 1-6

Algebra of Sets and Duality

Sets under the operations of union, intersection, and complement satisfy various laws or identities which are listed in Table 1-1. In fact, we formally state this:

Theorem 1.3: Sets satisfy the laws in Table 1-1.

A method of proving equations involving set operations is to use what it means for an object x to be an element of each side. Venn diagrams may be used as a guide for developing such an argument. A sec-

Table 1-1 Laws of the algebra of sets

Idempotent laws	
(1a) $A \cup A = A$	(1b) $A \cap A = A$
Associative laws	
(2a) $(A \cup B) \cup C = A \cup (B \cup C)$	(2b) $(A \cap B) \cap C = A \cap (B \cap C)$
Commutative laws	
(3a) $A \cup B = B \cup A$	(3b) $A \cap B = B \cap A$
Distributive laws	
(4a) $A \cup (B \cap C) = (A \cup B) \cap (A \cup C)$	(4b) $A \cap (B \cup C) = (A \cap B) \cup (A \cap C)$
Identity laws	
(5a) $A \cup \varnothing = A$	(5b) $A \cap U = A$
(6a) $A \cup U = U$	(6b) $A \cap \varnothing = \varnothing$
Involution laws	
(7) $(A^c)^c = A$	
Complement laws	
(8a) $A \cup A^c = U$	(8b) $A \cap A^c = \varnothing$
(9a) $U^c = \varnothing$	(9b) $\varnothing^c = U$
DeMorgan's laws	
(10a) $(A \cup B)^c = A^c \cap B^c$	(10b) $(A \cap B)^c = A^c \cup B^c$

ond method of proof is the use of identities, e.g., the following theorem $(A \backslash B)^c = A^c \cup B$ can be proved by:

$$(A \backslash B)^c = \left(A \cap B^c \right)^c$$
$$= A^c \cup B^{cc}$$
$$= A^c \cup B$$

Duality

Note that the identities in Table 1-1 are arranged in pairs as, for example, (2a) and (2b). We now consider the principle behind this arrangement. Suppose E is an equation of set algebra. The *dual* E^* of E is the equation obtained by replacing each occurrence of $\cup$, $\cap$, U, and $\varnothing$ in E by $\cap$, $\cup$, $\varnothing$ and U, respectively. Observe that the pairs of laws in Table 1-1 are duals of each other. It is a fact of set algebra, called the *principle of duality*, that, if any equation E is an identity, then its dual E^* is also an identity.

Finite Sets, Counting Principle

A set is said to be finite if it contains exactly m distinct elements where m denotes some nonnegative integer. Otherwise, a set is said to be infinite. For example, the empty set $\varnothing$ and the set of letters of the English alphabet are finite sets, whereas the set of even positive integers, $\{2, 4, 6, \dots \}$, is infinite.

The notation $n(A)$ will denote the number of elements in a finite set A.

Lemma 1.4: If A and B are disjoint finite sets, then $A \cup B$ is finite and

$$n(A \cup B) = n(A) + n(B)$$

Proof. In counting the elements of $A \cup B$, first count those that are in A. There are $n(A)$ of these. The only other elements of $A \cup B$ are those that are in B but not in A. But since A and B are disjoint, no element of B is in A, so there are $n(B)$ elements that are in B but not in A. Therefore, $n(A \cup B) = n(A) + n(B)$.

Theorem 1.5: If A and B are finite sets, then $A \cup B \cup C$, and

$$n(A \cup B) = n(A) + n(B) - n(A \cap B)$$

We can apply this result to obtain a similar formula for three sets:

Corollary 1.6 If A, B, and C are finite sets, then so is $A \cup B \cup C$, and

$$n(A \cup B \cup C) = n(A) + n(B) + n(C)$$
$$- n(A \cap B) - n(A \cap C) - n(B \cap C) + n(A \cap B \cap C)$$

Classes of Sets, Power Sets, Partitions

Given a set S, we might wish to talk about some of its subsets. Thus, we would be considering a set of sets. Whenever such a situation occurs, to avoid confusion we will speak of a *class* of sets or *collection* of sets rather than a set of sets. If we wish to consider some of the sets in a given class of sets, then we speak of a *subclass* or *subcollection*.

Power Sets

For a given set S, we may speak of the class of all subsets of S. This class is called the *power set* of S, and will be denoted by Power(S). If S is finite, then so is Power(S). In fact, the number of elements in Power(S) is 2 raised to the power of S; that is

$$n(\text{Power}(S)) = 2^{n(S)}$$

Partitions

Let S be a nonempty set. A partition of S is a subdivision of S into nonoverlapping, nonempty subsets. Precisely, a *partition* of S is a collection $\{A_1\}$ of nonempty subsets of S such that:
  (i) Each a in S belongs to one of the A_i.
  (ii) The sets of $\{A_i\}$ are mutually disjoint; that is, if

$$A_i \neq A_j \text{ then } A_i \cap A_j = \varnothing$$

The subsets in a partition are called *cells*. Figure 1-7 is a Venn diagram of a partition of the rectangular set S of points into five cells, $A_1, A_2, A_3, A_4,$ and A_5.

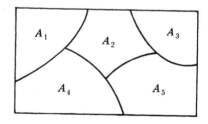

Figure 1-7

Generalized Set Operations

The set operations of union and intersection were defined above for two sets. These operations can be extended to any number of sets, finite or infinite, as follows.

Consider first a finite number of sets, say, $A_1, A_2, \ldots, A_m$. The union and intersection of these sets are denoted and defined, respectively, by

$$A_1 \cup A_2 \cup L \cup A_m = \cup_{i=1}^{m} A_i = \{x : x \in A_i \text{ for some } A_i\}$$

$$A_1 \cap A_2 \cap L \cap A_m = \cap_{i=1}^{m} A_i = \{x : x \in A_i \text{ for every } A_i\}$$

That is, the union consists of those elements that belong to at least one of the sets, and the intersection consists of those elements that belong to all of the sets.

Now let $\mathbf{A}$ be any collection of sets. The union and intersection of the sets in the collection $\mathbf{A}$ is denoted and defined, respectively, by

$$\cup(A : A \in \mathbf{A}) = \{x : x \in A \text{ for some } A \in \mathbf{A}\}$$

$$\cap(A : A \in \mathbf{A}) = \{x : x \in A \text{ for every } A \in \mathbf{A}\}$$

That is, the union consists of those elements which belong to at least one of the sets in the collection $\mathbf{A}$, and the intersection consists of those elements which belong to every set in the collection $\mathbf{A}$.

Chapter 2
FUNCTIONS AND ALGORITHMS

Functions

Suppose that to each element of a set A, we assign a unique element of a set B; the collection of such assignments is called a *function* from A into B. The set A is called the *domain* of the function, and the set B is called the *codomain*.

Functions are ordinarily denoted by symbols. For example, let f denote a function from A into B. Then we write

$$f : A \to B$$

which is read: "f is a function from A into B," or "f takes (or maps) A into B." If $a \in A$, then $f(a)$ (read: "f of a") denotes the unique element of B which f assigns to a; it is called the *image* of a under f, or the *value* of f at a. The set of all image values is called the *range* or *image* of f. The image of $f : A \to B$ is denoted by Ran (f), Im (f) or $f(A)$.

Frequently, a function can be expressed by means of a mathematical formula. For example, consider the function that sends each real number into its square. We may describe this function by writing

$$f(x) = x^2 \quad \text{or} \quad y = x^2$$

In the first notation, x is called a *variable* and the letter f denotes the function. In the other notation, x is called the *independent variable* and y is called the *dependent variable* since the value of y will depend on the value of x.

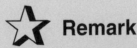 **Remark**

Whenever a function is given by a formula in terms of a variable x, we assume, unless it is otherwise stated, that the domain of the function is **R** (or the largest subset of **R** for which the formula has meaning) and the codomain is **R**.

Composite Function

Consider functions $f : A \to B$ and $g : B \to C$; that is, where the codomain of f is the domain of g. Then we may define a new function from A to C, called the *composition* of f and g and written $g \circ f$, as follows:

$$(g \circ f)(a) \equiv g(f(a))$$

That is, we find the image of a under f and then find the image of $f(a)$ under g. If we let A be any set, the function from A into A which assigns to each element that element itself is called the *identity function* on A and is usually denoted by 1_A or simply 1. In other words,

$$1_A(a) = a$$

Consider any function $f: A \rightarrow B$. Then

$$f \circ 1_A = f \qquad \text{and} \qquad 1_B \circ f = f$$

where 1_A and 1_B are the identity functions on A and B, respectively.

One-to-One, Onto, and Invertible Functions

A function $f: A \rightarrow B$ is said to be *one-to-one* (also written 1-1) if different elements in the domain A have distinct images. Another way of saying the same thing is that f is *one-to-one* if $f(a) = f(a')$ implies $a = a'$.

A function $f: A \rightarrow B$ is said to be an *onto* function if each element of B is the image of some element of A. In other words, $f: A \rightarrow B$ is onto if the image of f is the entire codomain, i.e., if $f(A) = B$. In such a case, we say that f is a function from A *onto* B or that f maps A *onto* B.

A function $f: A \rightarrow B$ is *invertible* if there is a function $g: B \rightarrow A$ such that $f \circ g = 1_B$ and $g \circ f = 1_A$. In general, there may not be such a function. But if there is one, then it is unique, it is denoted by f^{-1}, and f is said to be invertible. The following theorem gives a simple criteria for invertibility.

Theorem 2.1: A function $f: A \rightarrow B$ is invertible if and only if f is both one-to-one and onto.

If $f: A \rightarrow B$ is one-to-one and onto, then f is called a *one-to-one correspondence* between A and B. This terminology comes from the fact that each element of A will then correspond to a unique element of B and vice versa. f^{-1} simply reverses the direction of this correspondence.

Mathematical Functions; Exponential and Logarithmic Functions

This section presents various mathematical functions that appear often in the analysis of algorithms together with their notation. We also discuss the exponential and logarithmic functions, and their relationship.

Floor and Ceiling Functions

Let x be any real number. Then x lies between two integers called the floor and the ceiling of x. Specifically,

> $\lfloor x \rfloor$, called the *floor* of x, denotes the greatest integer that does not exceed x.
> $\lceil x \rceil$, called the *ceiling* of x, denotes the least integer that is not less than x.

If x is itself an integer, then $\lfloor x \rfloor = \lceil x \rceil$; otherwise $\lfloor x \rfloor + 1 = \lceil x \rceil$.

Integer and Absolute Value Functions

Let x be any real number. The *integer value* of x, written $INT(x)$, converts x into an integer by deleting (truncating) the fractional part of the number. Thus,

$$INT(3.14) = 3, \quad INT(\sqrt{5}) = 2, \quad INT(-8.5) = -8$$

Observe that $INT(x) = \lfloor x \rfloor$ or $INT(x) = \lceil x \rceil$ according to whether x is positive or negative.

The *absolute value* of the real number x, written $ABS(x)$ or $|x|$, is defined as the greater of x or $-x$. Hence $ABS(0) = 0$, and, for $x \neq 0$, $ABS(x) = x$ or $ABS(x) = -x$, depending on whether x is positive or negative. Thus,

$$|-15| = 15, \qquad |7| = 7, \qquad |-3.33| = 3.33$$

We note that $|x| = |-x|$ and, for $x \neq 0, |x|$ is positive.

Remainder Function; Modular Arithmetic

Let k be any integer and let M be a positive integer. Then,

$$k \ (\text{mod } M)$$

(read k *modulo* M) will denote the integer remainder when k is divided by M. More exactly, $k \ (\text{mod } M)$ is the unique integer r such that

$$k = Mq + r \quad \text{where} \quad 0 \le r < M$$

When k is positive, simply divide k by M to obtain the remainder r. Thus,

$$25 \ (\text{mod } 7) = 4, \quad 25 \ (\text{mod } 5) = 0, \quad 35 \ (\text{mod } 11) = 2$$

If k is negative, divide $|k|$ by M to obtain a remainder r'; then $k \ (\text{mod } M) = M - r'$ when $r' \ne 0$. Thus,

$$-26 \ (\text{mod } 7) = 7 - 5 = 2, \quad -371 \ (\text{mod } 8) = 8 - 3 = 5, \quad -39 \ (\text{mod } 3) = 0$$

The term "mod" is also used for the mathematical congruence relation, which is denoted and defined as follows:

$$a \equiv b \ (\text{mod } M) \quad \text{if and only if} \quad M \text{ divides } b - a$$

M is called the *modulus*, and $a \equiv b \ (\text{mod } M)$ is read "a is congruent to b modulo M." The following aspects of the congruence relation are frequently useful:

$$0 \equiv M \ (\text{mod } M) \quad \text{and} \quad a \pm M \equiv a \ (\text{mod } M)$$

Arithmetic modulo M refers to the arithmetic operations of addition, multiplication, and subtraction where the arithmetic value is replaced by its equivalent value in the set

$$\{0, 1, 2, \ldots, M - 1\}$$

or in the set

$$\{1,2,3,\ldots,M\}$$

Exponential Functions

Recall the following definitions for integer exponents (where m is a positive integer):

$$a^m = a \cdot a \cdots a \ (m \text{ times}), \qquad a^0 = 1, \qquad a^{-m} = \frac{1}{a^m}$$

Exponents are extended to include all rational numbers by defining, for any rational number m/n,

$$a^{m/n} = \left(\sqrt[n]{a}\right)^m$$

When $a > 0$, the exponential function base a $f(x) = a^x$ may be extended to all real numbers, x by a limiting process: $a^x = \lim a^r, r \to x$ where r is a rational number.

Logarithmic Functions

Logarithms are related to exponents as follows. Let b be a positive number. The logarithm of any positive number x to be the base b, written

$$\log_b x$$

represents the exponent to which b must be raised to obtain x. That is,

$$y = \log_b x \quad \text{and} \quad b^y = x$$

are equivalent statements. These statements mean that $\log_b$ is the inverse of the base b exponential function. Accordingly,

$\log_2 8 = 3 \quad$ since $\quad 2^3 = 8; \quad \log_{10} 100 = 2 \quad$ since $\quad 10^2 = 100$

$\log_2 64 = 6 \quad$ since $\quad 2^6 = 64; \quad \log_{10} 0.001 = -3 \quad$ since $\quad 10^{-3} = 0.001$

Furthermore, for any base b,

$$\log_b 1 = 0 \quad \text{since} \quad b^0 = 1; \quad \log_b b = 1 \quad \text{since} \quad b^1 = b$$

The logarithm of a negative number and the logarithm of 0 are not defined.

Sequences, Indexed Classes of Sets

Sequences and indexed classes of sets are special types of functions with their own notation. We discuss these objects in this section. We also discuss the summation notation here.

Sequences

A *sequence* is a function from the set $\mathbf{N} = \{1, 2, 3, \ldots\}$ of positive integers into a set A. The notation a_n is used to denote the image of the integer n. Thus, a sequence is usually denoted by

$$a_1, a_2, a_3, \ldots \quad \text{or} \quad \{a_n : n \in \mathbf{N}\} \quad \text{or simply} \quad \{a_n\}$$

Sometimes the domain of a sequence is the set $\{0, 1, 2, \ldots\}$ of nonnegative integers rather than $\mathbf{N}$. In such a case, we say n begins with 0 rather than 1.

A *finite sequence* over a set A is a function from $\{1, 2, \ldots, m\}$ into A, and it is usually denoted by $a_1, a_2, a_3, \ldots, a_m$. Such a finite sequence is sometimes called a *list* or an *m-tuple*.

Summation Symbol, Sums

Here we introduce the summation symbol Σ (the Greek letter sigma). Consider a sequence $a_1, a_2, a_3, \ldots$ Then the sums

$$a_1 + a_2 + \cdots + a_n \quad \text{and} \quad a_m + a_{m+1} + \cdots + a_n$$

will be denoted, respectively, by

$$\sum_{j=1}^{n} a_j \quad \text{and} \quad \sum_{j=m}^{n} a_j$$

The letter j in the above expressions is called a *dummy index* or *dummy variable.*

Indexed Classes of Sets

Let I be any nonempty set, and let S be a collection of sets. An *indexing function* from I to S is a function $f: I \to S$. For any $i \in I$, if we denote the image $f(i)$ by A_i, then the indexing function f may be denoted by

$$\{A_i : i \in I\} \qquad \text{or} \qquad \{A_i\}_{i \in I} \qquad \text{or simply} \qquad \{A_i\}$$

The set I is called the *indexing set*, and the elements of I are called *indices*. If f is one-to-one and onto, we say that S is indexed by I.

Recursively Defined Functions

A function defined on a set of integers is said to be *recursively defined* if the function definition refers to itself. In order for the definition not to be circular, the function definition must have the following two properties:

1. There must be certain arguments, called *base values*, for which the function does not refer to itself.
2. Each time the function does refer to itself, the argument of the function must be closer to a base value.

A recursive function with these two properties is said to be *well-defined.* The following examples should help clarify these ideas.

Factorial Function

The product of the positive integers from 1 to n, inclusive, is called "n factorial" and is usually denoted by $n!$; that is

$$n! = 1 \cdot 2 \cdot 3 \cdots (n-2)(n-1)n$$

It is also convenient to define $0! = 1$, so that the function is defined for all nonnegative integers. Thus we have

$$0! = 1, \quad 1! = 1, \quad 2! = 2 \cdot 1 = 2, \quad 3! = 3 \cdot 2 \cdot 1 = 6$$

$$4! = 4 \cdot 3 \cdot 2 \cdot 1 = 24, \quad 5! = 5 \cdot 4 \cdot 3 \cdot 2 \cdot 1 = 120$$

and so on. Observe that

$$4! = 4 \cdot 3! = 4 \cdot 6 = 24 \quad \text{and} \quad 5! = 5 \cdot 4! = 5 \cdot 24 = 120$$

This is true for every positive integer n; that is,

$$n! = n \cdot (n - 1)!$$

Accordingly, the factorial function may also be defined as follows:

Definition 2.1 (Factorial Function):

    (*a*) If $n = 0$, then $n! = 1$.
    (*b*) If $n > 0$, then $n! = n \cdot (n - 1)!$

Observe that the above definition of $n!$ is recursive, since it refers to it-self when it uses $(n - 1)!$. However,

1. The value of $n!$ is explicitly given when $n = 0$ (thus 0 is a base value).
2. The value of $n!$ for arbitrary n is defined in terms of a smaller value of n which is closer to the base value 0.

Accordingly, the factorial function is well-defined.

Fibonacci Sequence

The celebrated Fibonacci sequence (usually denoted by $F_0, F_1, F_2, \ldots$) is as follows:

$$0, 1, 1, 2, 3, 5, 8, 13, 21, 34, 55, \ldots$$

That is, $F_0 = 0$, $F_1 = 1$ and each succeeding term is the sum of the two preceding terms. For example, the next two terms of the sequence are

$$34 + 55 = 89 \quad \text{and} \quad 55 + 89 = 144$$

A formal definition of this function follows:

Definition 2.2 (Fibonacci Sequence):

    (*a*) If $n = 0$ or $n = 1$, then $F_n = n$.
    (*b*) If $n > 1$, then $F_n = F_{n-2} + F_{n-1}$.

This is another example of a recursive definition, since the definition refers to itself when it uses F_{n-2} and F_{n-1}. However,

1. The base values are 0 and 1.
2. The value of F_n is defined in terms of smaller values of n which are closer to the base values.

Accordingly, this function is well-defined.

Ackermann Function

The Ackermann function is a function with two arguments, each of which can be assigned any nonnegative integer, that is, 0, 1, 2, ... This function is defined as follows:

Definition 2.3 (Ackermann Function):

    (*a*) If $m = 0$, then $A(m, n) = n + 1$.
    (*b*) If $m \neq 0$ but $n = 0$, then $A(m, n) = A(m-1, 1)$.
    (*c*) If $m \neq 0$ and $n \neq 0$, then $A(m, n) = A(m-1, A(m, n-1))$.

Once more, we have a recursive definition, since the definition refers to itself in parts (*b*) and (*c*). Observe that $A(m, n)$ is explicitly given only when $m = 0$. The base criteria are the pairs

$$(0,0), (0,1), (0,2), (0,3),\ldots, (0,n), \ldots$$

Although it is not obvious from the definition, the value of any $A(m, n)$ may eventually be expressed in terms of the value of the function on one or more of the base pairs.

Algorithms and Functions

An algorithm M is a finite step-by-step list of well-defined instructions for solving a particular problem, say, to find the output $f(X)$ for a given function f with input X. (Here X may be a list or set of values.) Frequently, there may be more than one way to obtain $f(X)$, as illustrated by the following examples.

Example 2.1 (Polynomial Evaluation) Suppose, for a given polynomial $f(x)$ and value $x = a$, we want to find $f(a)$, say

$$f(x) = 2x^3 - 7x^2 + 4x - 15 \quad \text{and} \quad a = 5$$

This can be done in the following two ways.

(a) (**Direct Method**): Here we substitute $a = 5$ directly in the polynomial to obtain

$$f(5) = 2(125) - 7(25) + 4(5) - 7 = 250 - 175 + 20 - 15 = 80$$

Observe that there are $3 + 2 + 1 = 6$ multiplications and 3 additions. In general, evaluating a polynomial of degree n directly would require approximately:

$$n + (n-1) + \cdots + 1 = \frac{n(n+1)}{2} \text{ multiplications and } n \text{ additions}$$

(b) (**Horner's Method**): Here we rewrite the polynomial by successively factoring out x (on the right) as follows:

$$f(x) = (2x^2 - 7x + 4)x - 15 = ((2x - 7)x + 4)x - 15$$

Then

$$f(5) = ((3)5 + 4)5 - 15 = (19)5 - 15 = 95 - 15 = 80$$

Observe that here there are 3 multiplications and 3 additions. In general, evaluating a polynomial of degree n by Horner's method would require approximately

$$n \text{ multiplications} \quad \text{and} \quad n \text{ additions}$$

Clearly, Horner's method is more efficient than the direct method.

Example 2.2 (Greatest Common Divisor) Let a and b be positive integers with, say, $b < a$; and suppose we want to find $d = \text{GCD}(a,b)$, the greatest common divisor of a and b. This can be done in the following two ways:

(*a*) (**Direct Method**): Here we find all the divisors of a, say by testing all the numbers from 2 to $a/2$, and all the divisors of b. Then we pick the largest common divisor. For example, suppose $a = 258$ and $b = 60$. The divisors of a and b follow:

$$a = 258; \quad \text{divisors:} \quad 1, 2, 3, 6, 86, 129, 258$$
$$b = 60; \quad \text{divisors:} \quad 1, 2, 3, 4, 5, 6, 10, 12, 15, 20, 30, 60$$

Accordingly, $d = \text{GCD}(258, 60) = 6$.

(*b*) (**Euclidean Algorithm**): Here we divide a by b to obtain a remainder r_1. (Note $r_1 < b$.) Then we divide b by the remainder r_1 to obtain a second remainder r_2. (Note $r_2 < r_1$.) Next we divide r_1 by r_2 to obtain a third remainder r_3. (Note $r_3 < r_2$.) We continue dividing r_k by r_{k+1} to obtain a remainder r_{k+2}. Since

$$a > b > r_1 > r_2 > r_3 \cdots$$

eventually we obtain a remainder $r_m = 0$. Then $r_{m-1} = \text{GCD}(a,b)$. For example, suppose $a = 258$ and $b = 60$. Then:

(1) Dividing $a = 258$ by $b = 60$ yields the remainder $r_1 = 18$.
(2) Dividing $b = 60$ by $r_1 = 18$ yields the remainder $r_2 = 6$.
(3) Dividing $r_1 = 18$ by $r_2 = 6$ yields the remainder $r_3 = 0$.

Thus, $r_2 = 6 = \text{GCD}(258, 60)$.

Solved Problem 2.1 Let A be the set of students in a school. Determine which of the following assignments defines a function on A.

(*a*) To each student assign his age.
(*b*) To each student assign his teacher.
(*c*) To each student assign his sex.
(*d*) To each student assign his spouse.

Solution. A collection of assignments is a function on A if and only if each element a in A is assigned exactly one element. Thus,

(*a*) Yes, because each student has one and only one age.
(*b*) Yes, if each student has only one teacher; no, if any student has more than one teacher.
(*c*) Yes.
(*d*) No, if any student is not married; yes, otherwise.

Solved Problem 2.2 Let the functions $f: A \to B$, $g: B \to C$, and $h: C \to D$ be defined by Figure 2-1. Determine if each function is onto.

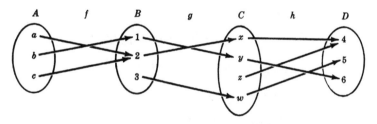

Figure 2-1

Solution.

The function $f: A \to B$ is not onto since $3 \in B$ is not the image of any element in A.

The function $g: B \to C$ is not onto since $z \in C$ is not the image of any element in B.

The function $h: C \to D$ is onto since each element in D is the image of some element of C.

Chapter 3
LOGIC AND PROPOSITIONAL CALCULUS

Propositions and Compound Propositions

A *proposition* (or *statement*) is a declarative sentence that is true or false, but not both. Consider, for example, the following eight sentences:

(i) Paris is in France. (v) $9 < 6$.
(ii) $1 + 1 = 2$. (vi) $x = 2$ is a solution of $x^2 = 4$.
(iii) $2 + 2 = 3$. (vii) Where are you going?
(iv) London is in Denmark. (viii) Do your homework.

All of them are propositions except (vii) and (viii). Moreover, (i), (ii), and (vi) are true, whereas (iii), (iv), and (v) are false.

Compound Propositions

Many propositions are *composite*, that is, composed of *subpropositions* and various connectives discussed subsequently. Such composite propositions are called *compound propositions*. A proposition is said to be *primitive* if it cannot be broken down into simpler propositions, that is, if it is not composite.

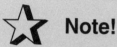 **Note!**

The fundamental property of a compound proposition is that its truth value is completely determined by the truth values of its subpropositions together with the way in which they are connected to form the compound propositions.

Basic Logical Operations

This section discusses the three basic logical operations of conjunction, disjunction, and negation which correspond, respectively, to the English words "and," "or," and "not."

Conjunction, $p \wedge q$

Any two propositions can be combined by the word "and" to form a compound proposition called the *conjunction* of the original propositions. Symbolically,

$$p \wedge q$$

read "p and q," denotes the conjunction of p and q. Since $p \wedge q$ is a proposition, it has a truth value, and this truth value depends only on the truth values of p and q. Specifically:

Definition 3.1: If p and q are true, then $p \wedge q$ is true; otherwise $p \wedge q$ is false.

The truth value of $p \wedge q$ may be defined equivalently by the table in Figure 3-1(a). Here, the first line is a short way of saying that if p is true and q is true, then $p \wedge q$ is true. The second line says that if p is true and q is false, then $p \wedge q$ is false. And so on. Observe that there are four lines corresponding to the four possible combinations of T and F for the two subpropositions p and q. Note that $p \wedge q$ is true only when both p and q are true.

p	q	$p \wedge q$
T	T	T
T	F	F
F	T	F
F	F	F

(a) "p and q"

p	q	$p \vee q$
T	T	T
T	F	T
F	T	T
F	F	F

(b) "p or q"

p	$\neg p$
T	F
F	T

(c) "not p"

Figure 3-1

Disjunction, $p \vee q$

Any two propositions can be combined by the word "or" to form a compound proposition called the *disjunction* of the original propositions. Symbolically,

$$p \vee q$$

read "*p* or *q*," denotes the disjunction of *p* and *q*. The truth value of $p \vee q$ depends only on the truth values of *p* and *q* as follows:

Definition 3.2: If *p* and *q* are false, then $p \vee q$ is false; otherwise $p \vee q$ is true.

The truth value of $p \vee q$ may be defined equivalently by the table in Figure 3-1(*b*). Observe that $p \vee q$ is false only in the fourth case when both *p* and *q* are false.

 Important!

The English word "or" is commonly used in two distinct ways. Sometimes it is used in the sense of "*p* or *q* or both," i.e., at least one of the two alternatives occurs, as above, and sometimes it is used in the sense of "*p* or *q* but not both," i.e., exactly one of the two alternatives occurs. Unless otherwise stated, "or" shall be used in the former sense.

Negation, ¬*p*

Given any proposition *p*, another proposition, called the *negation* of *p*, can be formed by writing "It is not the case that…" or "It is false that…" before *p* or, if possible, by inserting in *p* the word "not." Symbolically,

$$\neg p$$

read "not *p*," denotes the negation of *p*. The truth value of $\neg p$ depends on the truth value of *p* as follows.

Definition 3.3: If *p* is true, then $\neg p$ is false; and if *p* is false, then $\neg p$ is true.

The truth value of $\neg p$ may be defined equivalently by the table in Figure 3-1(c). Thus, the truth value of the negation of p is always the opposite of the truth value of p.

Propositions and Truth Tables

Let $P(p, q, \ldots)$ denote an expression constructed from logical variables $p, q, \ldots$, which take on the value TRUE (T) or FALSE (F), and the logical connectives $\land$, $\lor$, and $\neg$ (and others discussed subsequently). Such an expression $P(p, q, \ldots)$ will be called a *proposition*.

The main property of a proposition $P(p, q, \ldots)$ is that its truth value depends exclusively upon the truth values of its variables, that is, the truth value of a proposition is known once the truth value of each of its variables is known. A simple, concise way to show this relationship is through a *truth table*. We describe a way to obtain such a truth table below.

Consider, for example, the proposition $\neg(p \land \neg q)$. Figure 3-2(a) indicates how the truth table of $\neg(p \land \neg q)$ is constructed. Observe that the first columns of the table are for the variables $p, q, \ldots$ and that there are enough rows in the table to allow for all possible combinations of T and F for these *variables*. (For 2 variables, as above, 4 rows are necessary; for 3 variables, 8 rows are necessary; and, in general, for n variables, 2^n rows are required.) There is then a column for each "elementary" stage of the construction of the proposition, the truth value at each step being determined from the previous stages by the definitions of the connectives $\land$, $\lor$, and $\neg$. Finally, we obtain the truth value of the proposition, which appears in the last column.

p	q	$\neg q$	$p \land \neg q$	$\neg(p \land \neg q)$
T	T	F	F	T
T	F	T	T	F
F	T	F	F	T
F	F	T	F	T

(a)

p	q	$\neg(p \land \neg q)$
T	T	T
T	F	F
F	T	T
F	F	T

(b)

Figure 3-2

In order to avoid an excessive number of parentheses, we sometimes adopt an order of precedence for the logical connectives. Specifically,

$\neg$ has precedence over $\wedge$ which has precedence over $\vee$

For example, $\neg p \wedge q$ means $(\neg p) \wedge q$ and not $\neg (p \wedge q)$.

Tautologies and Contradictions

Some propositions $P(p, q, \ldots)$ contain only T in the last column of their truth tables or, in other words, they are true for any truth values of their variables. Such propositions are called *tautologies*. Analogously, a proposition $P(p, q, \ldots)$ is called a *contradiction* if it contains only F in the last column of its truth table or, in other words, if it is false for any truth values of its variables. For example, the proposition "p or not p," that is, $p \vee \neg p$, is a tautology, and the proposition "p and not p," that is, $p \wedge \neg p$, is a contradiction. This is verified by looking at their truth tables in Figure 3-3. (The truth tables have only two rows since each proposition has only the one variable p.)

p	$\neg p$	$p \vee \neg p$
T	F	T
F	T	T

(a) $p \vee \neg p$

p	$\neg p$	$p \wedge \neg p$
T	F	F
F	T	F

(b) $p \wedge \neg p$

Figure 3-3

Note that the negation of a tautology is a contradiction since it is always false, and the negation of a contradiction is a tautology since it is always true.

Now let $P(p, q, \ldots)$ be a tautology, and let $P_1(p, q, \ldots)$, $P_2(p, q, \ldots)$,…be any propositions. Since $P(p, q, \ldots)$ does not depend upon the particular truth values of its variables $p, q, \ldots$, we can substitute P_1 for p, P_2 for $q, \ldots$ in the tautology $P(p, q, \ldots)$ and still have a tautology. In other words:

Theorem 3.1 (Principle of Substitution): If $P(p, q, \ldots)$ is a tautology, then $P(P_1, P_2, \ldots)$ is a tautology for any propositions $P_1, P_2, \ldots$.

Logical Equivalence

Two propositions $P(p, q, \ldots)$ and $Q(p, q, \ldots)$ are said to be *logically equivalent*, or simply *equivalent* or *equal*, denoted by

$$P(p, q, \ldots) \equiv Q(p, q, \ldots)$$

if they have identical truth tables. Consider, for example, the truth tables of $\neg (p \wedge q)$ and $\neg p \vee \neg q$ appearing in Figure 3-4. Observe that both truth tables are the same, that is, both propositions are false in the first case and true in the other three cases. Accordingly, we can write

$$(\neg p \wedge q) \equiv \neg p \vee \neg q$$

In other words, the propositions are logically equivalent.

p	q	$p \wedge q$	$\neg(p \wedge q)$
T	T	T	F
T	F	F	T
F	T	F	T
F	F	F	T

(a) $\neg(p \wedge q)$

p	q	$\neg p$	$\neg q$	$\neg p \vee \neg q$
T	T	F	F	F
T	F	F	T	T
F	T	T	F	T
F	F	T	T	T

(b) $\neg p \vee \neg q$

Figure 3-4

The actual truth table of the proposition $\neg (p \wedge \neg q)$ is shown in Figure 3-4(b). It consists precisely of the columns in Figure 3-4(a) which appear under the variables and under the proposition; the other columns were merely used in the construction of the truth table.

Algebra of Propositions

Propositions satisfy various laws which are listed in Table 3-1. (In this table, T and F are restricted to the truth values "true" and "false," respectively.) We state this result formally.

Theorem 3.2: Propositions satisfy the laws of Table 3-1.

Table 3-1   Laws of the algebra of propositions

Idempotent laws	
(1a) $p \vee p \equiv p$	(1b) $p \wedge p \equiv p$
Associative laws	
(2a) $(p \vee q) \vee r \equiv p \vee (q \vee r)$	(2b) $(p \wedge q) \wedge r \equiv p \wedge (q \wedge r)$
Commutative laws	
(3a) $p \vee q \equiv q \vee p$	(3b) $p \wedge q \equiv q \wedge p$
Distributive laws	
(4a) $p \vee (q \wedge r) \equiv (p \vee q) \wedge (p \vee r)$	(4b) $p \wedge (q \vee r) \equiv (p \wedge q) \vee (p \wedge r)$
Identity laws	
(5a) $p \vee F \equiv p$	(5b) $p \wedge T \equiv p$
(6a) $p \vee T \equiv T$	(6b) $p \wedge F \equiv F$
Complement laws	
(7a) $p \vee \neg p \equiv T$	(7b) $p \wedge \neg p \equiv F$
(8a) $\neg T \equiv F$	(8b) $\neg F \equiv T$
Involution law	
(9) $\neg \neg p \equiv p$	
DeMorgan's laws	
(10a) $\neg (p \vee q) \equiv \neg p \wedge \neg q$	(10b) $\neg (p \wedge q) \equiv \neg p \vee \neg q$

Conditional and Biconditional Statements

Many statements, particularly in mathematics, are of the form "If p then q." Such statements are called *conditional* statements and are denoted by

$$p \rightarrow q$$

The conditional $p \rightarrow q$ is frequently read "p implies q" or "p only if q."
Another common statement is of the form "p if and only if q." Such statements are called *biconditional* statements and are denoted by

$$p \leftrightarrow q$$

The truth values of $p \rightarrow q$ and $p \leftrightarrow q$ are defined by the tables in Figure 3-5. Observe that:

(a) The conditional $p \to q$ is false only when the first part p is true and the second part q is false. Accordingly, when p is false, the conditional $p \to q$ is true regardless of the truth value of q.

(b) The biconditional $p \leftrightarrow q$ is true whenever p and q have the same truth values and false otherwise.

p	q	$p \to q$
T	T	T
T	F	F
F	T	T
F	F	T

(a) $p \to q$

p	q	$p \leftrightarrow q$
T	T	T
T	F	F
F	T	F
F	F	T

(b) $p \leftrightarrow q$

Figure 3-5

The truth table of the proposition $\neg p \vee q$ appears in Figure 3-6. Observe that the truth tables of $\neg p \vee q$ and $p \to q$ have the same last columns, that is, they are both false only in the second case. Accordingly, $p \to q$ is logically equivalent to $\neg p \vee q$; that is,

$$p \to q \equiv \neg p \vee q$$

p	q	$\neg p$	$\neg p \vee q$
T	T	F	T
T	F	F	F
F	T	T	T
F	F	T	T

$\neg p \vee q$

Figure 3-6

In other words, the conditional statement "If p then q" is logically equivalent to the statement "Not p or q" which only involves the connectives $\vee$ and $\neg$ and thus was already a part of our language. We may regard $p \to q$ as an abbreviation for an oft-recurring statement.

Arguments

An *argument* is an assertion that a given set of propositions $P_1, P_2, ..., P_n$, called *premises*, yields (has a consequence) another proposition Q, called the *conclusion*. Such an argument is denoted by

$$P_1, P_2, ..., P_n \vdash Q$$

The notion of a "logical argument" or "*valid* argument" is formalized as follows:

Definition 3.4: An argument $P_1, P_2, ..., P_n \vdash Q$ is said to be valid if Q is true whenever all the premises $P_1, P_2, ..., P_n$ are true.

Remember

An argument that is not valid is called a *fallacy.*

Now the propositions $P_1, P_2, ..., P_n$ are true simultaneously if and only if the proposition $P_1 \wedge P_2 \wedge ... \wedge P_n$ is true. Thus the argument $P_1, P_2, ..., P_n \vdash Q$ is valid if and only if Q is true whenever $P_1 \wedge P_2 \wedge ... \wedge P_n$ is true or, equivalently, if the proposition $(P_1 \wedge P_2 \wedge ... \wedge P_n) \rightarrow Q$ is a tautology. We state this result formally.

Theorem 3.3: The argument $P_1, P_2, ..., P_n \vdash Q$ is valid if and only if the proposition $(P_1 \wedge P_2 \wedge ... \wedge P_n) \rightarrow Q$ is a tautology.

Propositional Functions, Quantifiers

Let A be a given set. A *propositional function* (or: an *open sentence* or *condition*) defined on A is an expression $p(x)$ which has the property that $p(a)$ is true or false for each $a \in A$. That is, $p(x)$ becomes a statement (with a truth value) whenever any element $a \in A$ is substituted for the variable x. The set A is called the *domain* of $p(x)$, and the set T_p of all elements of A for which $p(a)$ is true is called the *truth set* of $p(x)$. In other words,

$$T_p = \{x : x \in A, p(x) \text{ is true}\} \quad \text{or} \quad T_p = \{x : p(x)\}$$

Frequently, when A is some set of numbers, the condition $p(x)$ has the form of an equation or inequality involving the variable x.

Universal Quantifier

Let $p(x)$ be a propositional function defined on a set A. Consider the expression

$$(\forall x \in A)p(x) \quad \text{or} \quad \forall x \, p(x) \tag{3.1}$$

which reads "For every x in A, $p(x)$ is a true statement" or simply, "For all x, $p(x)$." The symbol $\forall$ which reads "for all" or "for every" is called the *universal quantifier*. The statement (3.1) is equivalent to the statement

$$T_p = \{x : x \in A, p(x)\} = A \tag{3.2}$$

that is, that the truth set of $p(x)$ is the entire set A.

The expression $p(x)$ by itself is an open sentence or condition and therefore has no truth value. However, $\forall x \, p(x)$, that is $p(x)$ preceded by the quantifier $\forall$, does have a truth value which follows from the equivalence of (3.1) and (3.2). Specifically:

Q_1: If $\{x : x \in A, p(x)\} = A$, then $\forall x \, p(x)$ is true; otherwise $\forall x \, p(x)$ is false.

Existential Quantifier

Let $p(x)$ be a propositional function defined on a set A. Consider the expression

$$(\exists x \in A)p(x) \quad \text{or} \quad \exists x, \, p(x) \tag{3.3}$$

which reads, "There exists an x in A such that $p(x)$ is a true statement" or simply, "For some x, $p(x)$." The symbol $\exists$ which reads "there exists" or, "for some" or "for at least one" is called the *existential quantifier*. Statement (3.3) is equivalent to the statement

$$T_p = \{x : x \in A, p(x)\} \neq \varnothing \tag{3.4}$$

i.e., that the truth set of $p(x)$ is not empty. Accordingly, $\exists x\, p(x)$, that is, $p(x)$ preceded by the quantifier $\exists$, does have a truth table. Specifically:

Q_2: If $\{x\colon p(x)\} \neq \varnothing$, then $\exists x\, p(x)$ is true; otherwise $\exists x\, p(x)$ is false.

Negation of Quantified Statements

Consider the statement: "All math majors are male." Its negation reads: "It is not the case that all math majors are male" or equivalently, "There exists at least one math major who is female (not male)." Symbolically, using M to denote the set of math majors, the above can be written as

$$\neg(\forall x \in M)(x \text{ is male}) \equiv (\exists x \in M)(x \text{ is not male})$$

or, when $p(x)$ denotes "x is male,"

$$\neg(\forall x \in M)p(x) \equiv (\exists x \in M)\neg p(x) \quad \text{or} \quad \neg\forall x\, p(x) \equiv \exists x\, \neg p(x)$$

The above is true for any proposition $p(x)$. That is:

Theorem 3.4 (DeMorgan): $\neg(\forall x \in A)p(x) \equiv (\exists x \in A)\neg p(x)$.

In other words, the following two statements are equivalent:

(1) It is not true that, for all $a \in A$, $p(a)$ is true.
(2) There exists an $a \in A$ such that $p(a)$ is false.

There is an analogous theorem for the negation of a proposition which contains the existential quantifier.

Theorem 3.5 (DeMorgan): $\neg(\exists x \in A)p(x) \equiv (\forall x \in A)\neg p(x)$.

That is, the following two statements are equivalent:

(1) It is not true that, for some $a \in A$, $p(a)$ is true.
(2) For all $a \in A$, $p(a)$ is false.

Propositional Functions with More Than One Variable

A propositional function (of n variables) defined over a product set (see Chapter 5) $A = A_1 \times A_2 \times \ldots \times A_n$ is an expression

$$p(x_1, x_2, \ldots, x_n)$$

which has the property that $p(a_1, a_2, \ldots, a_n)$ is true or false for any n-tuple $(a_1, \ldots, a_n)$ in A.

Basic Principle: A propositional function preceded by a quantifier for each variable, for example,

$$\forall x \exists y, p(x, y) \quad \text{or} \quad \exists x \forall y \exists z, p(x, y, z)$$

denotes a statement and has a truth value.

Negating Quantified Statements with More Than One Variable

Quantified statements with more than one variable may be negated by successively applying Theorems 3.4 and 3.5. Thus each $\forall$ is changed to $\exists$ and each $\exists$ is changed to $\forall$ as the negation symbol $\neg$ passes through the statement from left to right. For example,

$$\neg\left[\forall x \exists y \exists z, p(x, y, z)\right] \equiv \exists x \neg\left[\exists y \exists z, p(x, y, z)\right] \equiv \exists x \forall y\left[\neg \exists z, p(x, y, z)\right]$$
$$\equiv \exists x \forall y \forall z, \neg p(x, y, z)$$

Mathematical Induction

An essential property of the set

$$\mathbf{N} = \{1, 2, 3, \ldots\}$$

which is used in many proofs, follows:

Principle of Mathematical Induction I:

Let P be a proposition defined on the positive integers $\mathbf{N}$, i.e., $P(n)$ is either true or false for each n in $\mathbf{N}$. Suppose P has the following two properties:

  (i) $P(1)$ is true.
  (ii) $P(n + 1)$ is true whenever $P(n)$ is true.

Then P is true for every positive integer.

There is a form of the principle of mathematical induction which is sometimes more convenient to use. Although it appears different, it is really equivalent to the principle of induction.

Principle of Mathematical Induction II:

Let P be a proposition defined on the positive integers $\mathbf{N}$ such that:

  (i) $P(1)$ is true.
  (ii) $P(n)$ is true whenever $P(k)$ is true for all $1 \le k < n$.

Then P is true for every positive integer.

Remark: Sometimes one wants to prove that a proposition P is true for the set of integers

$$\{a, a + 1, a + 2, \dots\}$$

where a is any integer, possibly zero. This can be done by simply replacing 1 by a in either of the above Principles of Mathematical Induction.

Solved Problem 3.1 Determine the truth value of each of the following statements:

  (a) $4 + 2 = 5$ and $6 + 3 = 9$ w(c) $4 + 5 = 9$ and $1 + 2 = 4$
  (b) $3 + 2 = 5$ and $6 + 1 = 7$ (d) $3 + 2 = 5$ and $4 + 7 = 11$

Solution. The statement "p and q" is true only when both substatements are true. Thus: (a) false; (b) true; (c) false; (d) true.

Solved Problem 3.2 Rewrite the following statements without using the conditional:

    (*a*) If it is cold, he wears a hat.
    (*b*) If productivity increases, then wages rise.

Solution. Recall that "If p then q" is equivalent to "Not p or q"; that is, $p \rightarrow q \equiv \neg p \vee q$. Hence,

    (*a*) It is not cold or he wears a hat.
    (*b*) Productivity does not increase or wages rise.

Solved Problem 3.3 Verify that the proposition $p \vee \neg(p \wedge q)$ is tautology.

Solution. Construct the truth table of $p \vee \neg(p \wedge q)$ as shown in Figure 3-7. Since the truth value of $p \vee \neg(p \wedge q)$ is T for all values of p and q, the proposition is a tautology.

p	q	$p \wedge q$	$\neg(p \wedge q)$	$p \vee \neg(p \wedge q)$
T	T	T	F	T
T	F	F	T	T
F	T	F	T	T
F	F	F	T	T

Figure 3-7

Chapter 4
COUNTING

IN THIS CHAPTER:

- ✔ *Basic Counting Principles*
- ✔ *Factorial Notation*
- ✔ *Binomial Coefficients*
- ✔ *Permutations*
- ✔ *Combinations*
- ✔ *The Pigeonhole Principle*
- ✔ *The Inclusion-Exclusion Principle*
- ✔ *Ordered and Unordered Partitions*

Basic Counting Principles

Combinatorial analysis, which includes the study of permutations, combinations, and partitions, is concerned with determining the number of logical possibilities of some event without necessarily identifying every case. There are two basic counting principles used throughout.

Sum Rule Principle: Suppose some event E can occur in m ways and a second event F can occur in n ways, and suppose both events cannot oc-

43

cur simultaneously. Then E or F can occur in $m + n$ ways. More generally, suppose an event E_1 can occur in n_1 ways, a second event E_2 can occur in n_2 ways, a third event E_3 can occur in n_3 ways,..., and suppose no two of the events can occur at the same time. Then one of the events can occur in $n_1 + n_2 + n_3 + \cdots$ ways.

Product Rule Principle: Suppose there is an event E that can occur in m ways and, independent of this event, there is a second event F that can occur in n ways. Then combinations of E and F can occur in mn ways. More generally, suppose an event E_1 can occur in n_1 ways, and following E_1, a second event E_2 can occur in n_2 ways, and, following E_2, a third event E_3 can occur in n_3 ways, and so on. Then all the events can occur in the order indicated in $n_1 \cdot n_2 \cdot n_3 \cdots$ ways.

There is a set theoretical interpretation of the above two counting principles. Specifically, suppose $n(A)$ denotes the number of elements in a set A. Then:

1. **Sum Rule Principle**: If A and B are disjoint sets, then

$$n(A \cup B = n(A) + n(B)$$

2. **Product Rule Principle**: Let $A \times B$ be the Cartesian product of sets A and B. Then

$$n(A \times B) = n(A) \cdot n(B)$$

Factorial Notation

The product of the positive integers from 1 to n inclusive is denoted by $n!$ (read "n factorial"):

$$n! = 1 \cdot 2 \cdot 3 \cdots (n-2)(n-1)n$$

In other words, $n!$ is defined by

$$1! = 1 \qquad \text{and} \qquad n! = n \cdot (n-1)!$$

It is also convenient to define $0! = 1$.

Binomial Coefficients

The symbol $\binom{n}{r}$, where r and n are positive integers with $r \leq n$, is defined as follows

$$\binom{n}{r} = \frac{n(n-1)(n-2)\cdots(n-r+1)}{1 \cdot 2 \cdot 3 \cdots (r-1)r}$$

and can also be written as

$$\binom{n}{r} = \frac{n(n-1)(n-2)\cdots(n-r+1)}{1 \cdot 2 \cdot 3 \cdots (r-1)r} = \frac{n!}{r!(n-r)!}$$

But $n - (n - r) = r$; hence we have the following important relation:

$$\binom{n}{n-r} = \binom{n}{r}$$

or, in other words, if $a + b = n$, then

$$\binom{n}{a} = \binom{n}{b}$$

Binomial Coefficients and Pascal's Triangle

The numbers $\binom{n}{r}$ are called the *binomial coefficients* since they appear as the coefficients in the expansion of $(a+b)^n$. Specifically, one can prove that

$$(a+b)^n = \sum_{k=0}^{n} \binom{n}{k} a^{n-k} b^k$$

$$(a + b)^0 = 1$$
$$(a + b)^1 = a + b$$
$$(a + b)^2 = a^2 + 2ab + b^2$$
$$(a + b)^3 = a^3 + 3a^2b + 3ab^2 + b^3$$
$$(a + b)^4 = a^4 + 4a^3b + 6a^2b^2 + 4ab^3 + b^4$$
$$(a + b)^5 = a^5 + 5a^4b + 10a^3b^2 + 10a^2b^3 + 5ab^4 + b^5$$
$$(a + b)^6 = a^6 + 6a^5b + 15a^4b^2 + 20a^3b^3 + 15a^2b^4 + 6ab^5 + b^6$$

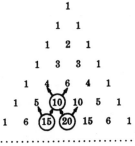

```
            1
          1   1
        1   2   1
      1   3   3   1
    1   4   6   4   1
  1   5  (10) 10  5   1
1   6 (15)(20) 15  6   1
```

Figure 4-1

The coefficients of the successive powers of $a + b$ can be arranged in a triangular array of numbers, called Pascal's triangle, as pictured in Figure 4-1.

The numbers in Pascal's triangle have the following intersecting properties:

(i) The first number and the last number in each row is 1.

(ii) Every other number in the array can be obtained by adding the two numbers appearing directly above it. For example, $10 = 4 + 6$, $15 = 5 + 10$, $20 = 10 + 10$.

Since the numbers appearing in Pascal's triangle are binomial coefficients, property (ii) of Pascal's triangle comes from the following theorem:

Theorem 4.1: $\begin{pmatrix} n+1 \\ r \end{pmatrix} = \begin{pmatrix} n \\ r-1 \end{pmatrix} + \begin{pmatrix} n \\ r \end{pmatrix}$

Permutations

Any arrangement of a set of n objects in a given order is called a *permutation* of the objects (taken all at a time). Any arrangement of any $r \le n$ of these objects in a given order is called an *r-permutation* or a *permutation of the n objects taken r at a time*. Consider, for example, the set of letters a, b, c, and d. Then:

- (i) *bdca*, *dcba*, and *acdb* are permutations of the four letters (taken all at a time);
- (ii) *bad*, *adb*, *cbd* and *bca* are permutations of the four letters taken three at a time;
- (iii) *ad*, *cb*, *da* and *bd* are permutations of the four letters taken two at a time.

The number of permutations of n objects taken r at a time is denoted by

$$P(n,r),\ _nP_r,\ P_{n,r},\ P_r^n,\ \text{or}\ (n)_r$$

We shall use $P(n,r)$, deriving an expression for the general formula for $P(n,r)$ below.

Derivation of the Formula for $P(n,r)$

The derivation of the formula for the number of permutations of n objects taken r at a time, or the number of r-permutations of n objects, $P(n,r)$, follows the procedure below. The first element in an r-permutation of n objects can be chosen in n different ways; following this, the second element in the permutation can be chosen in $n - 1$ ways; and, following this, the third element in the permutation can be chosen in $n - 2$ ways. Continuing in this manner, we have that the rth (last) element in the r-permutation can be chosen in $n - (r - 1) = n - r + 1$ ways. Thus, by the fundamental principle of counting, we have

$$P(n,r) = n(n-1)(n-2)\cdots(n-r+1)$$

It can also be shown that

$$n(n-1)(n-2)\cdots(n-r+1) = \frac{n(n-1)(n-2)\cdots(n-r+1)\cdot(n-r)!}{(n-r)!}$$

$$= \frac{n!}{(n-r)!}$$

Thus, we have proven the following theorem:

Theorem 4.2: $P(n,r) = \dfrac{n!}{(n-r)!}$

Corollary 4.3: There are $n!$ permutations of n objects (taken all at a time).

For example, there are $3! = 1 \cdot 2 \cdot 3 = 6$ permutations of the three letters a, b, and c. These are abc, acb, bac, bca, cab, cba.

Permutations with Repetitions

Frequently, we want to know the number of permutations of a multiset, that is, a set of objects some of which are alike. We will let

$$P(n;n_1,n_2,\ldots,n_r)$$

denote the number of permutations of n objects of which n_1 are alike, n_2 are alike, ..., n_r are alike. The general formula follows:

Theorem 4.4: $P(n;n_1,n_2,\ldots,n_r) = \dfrac{n!}{n_1!n_2!\cdots n_r!}$

Combinations

Suppose we have a collection of n objects. A *combination* of these n objects taken r at a time is any selection of r of the objects where order does not count. In other words, an *r-combination* of a set of n objects is any subset of r elements. For example, the combinations of the letters a, b, c, d taken three at a time are

$\{a,b,c\}, \{a,b,d\}, \{a,c,d\}, \{b,c,d\},$ or simply abc, abd, acd, bcd

Observe that the following combinations are equal:

$$abc, acb, bac, bca, cab, \text{ and } cba$$

That is, each denotes the same set $\{a, b, c\}$.

The number of combinations of n objects taken r at a time is denoted by $C(n, r)$. The symbols, $_nC_r$, $C_{n,r}$, and C_r^n also appear in various texts. The general formula for $C(n, r)$ will be given below.

Formula for $C(n, r)$

Since any combination of n objects taken r at a time determines $r!$ permutations of the objects in the combination, we can conclude that

$$P(n, r) = r!C(n, r)$$

Thus we obtain

Theorem 4.5: $C(n,r) = \dfrac{P(n,r)}{r!} = \dfrac{n!}{r!(n-r)!}$

Recall that the binomial coefficient $\dbinom{n}{r}$ was defined to be $\dfrac{n!}{r!(n-r)!}$; hence $C(n,r) = \dbinom{n}{r}$.

The Pigeonhole Principle

Many results in combinatorial theory come from the following almost obvious statement.

Pigeonhole Principle: If n pigeonholes are occupied by $n + 1$ or more pigeons, then at least one pigeonhole is occupied by more than one pigeon.

The principle can be applied to many problems where we want to show that a given situation can occur. For example, suppose a department contains 13 professors. Then two of the professors (pigeons) were born in the same month (pigeonholes).

The Pigeonhole Principle is generalized as follows:

Generalized Pigeonhole Principle: If n pigeonholes are occupied by $kn + 1$ or more pigeons, where k is a positive integer, then at least one pigeonhole is occupied by $k + 1$ or more pigeons.

The Inclusion-Exclusion Principle

Let A and B be any finite sets. Then

$$n(A \cup B) = n(A) + n(B) - n(A \cap B)$$

In other words, to find the number $n(A \cup B)$ of elements in the union $A \cup B$, we add $n(A)$ and $n(B)$ and then we subtract $n(A \cap B)$; that is, we "include" $n(A)$ and $n(B)$, and we "exclude" $n(A \cap B)$. This principle holds for any number of sets. We state it for three sets.

Theorem 4.6: For any finite sets A, B, C, we have

$$n(A \cup B \cup C) = n(A) + n(B) + n(C)$$
$$- n(A \cap B) - n(A \cap C) - n(B \cap C) + n(A \cap B \cap C)$$

Ordered and Unordered Partitions

Suppose a bag A contains seven marbles numbered 1 through 7. We compute the number of ways we can draw, first, two marbles from the bag, then three marbles from the bag, and lastly two marbles from the bag. In other words, we want to compute the number of *ordered partitions*

$$[A_1, A_2, A_3]$$

of the set of seven marbles into cells A_1 containing two marbles, A_2 containing three marbles and A_3 containing two marbles. We call these ordered partitions since we distinguish between

$$[\{1,2\}, \{3,4,5\}, \{6,7\}] \quad \text{and} \quad [\{6,7\}, \{3,4,5\}, \{1,2\}]$$

each of which determines the same partition of A.

Now we begin with seven marbles in the bag, so there are $\binom{7}{2}$ ways of drawing the first two marbles, i.e., of determining the first cell A_1; following this, there are five marbles left in the bag and so there are $\binom{5}{3}$ ways of drawing the three marbles, i.e., of determining the second cell A_2; finally, there are two marbles left in the bag and so there are $\binom{2}{2}$ ways of determining the last cell A_3. Hence, there are

$$\binom{7}{2}\binom{5}{3}\binom{2}{2} = \frac{7\cdot 6}{1\cdot 2}\cdot\frac{5\cdot 4\cdot 3}{1\cdot 2\cdot 3}\cdot\frac{2\cdot 1}{1\cdot 2} = 210$$

different ordered partitions of A into cells A_1 containing two marbles, A_2 containing three marbles, and A_3 containing two marbles.

Now observe that

$$\binom{7}{2}\binom{5}{3}\binom{2}{2} = \frac{7!}{2!5!}\cdot\frac{5!}{3!2!}\cdot\frac{2!}{2!0!} = \frac{7!}{2!3!2!}$$

since each numerator after the first is canceled by the second term in the denominator of the previous factor. The above discussion can be shown to hold in general. Namely,

Theorem 4.7: Let A contain n elements and let $n_1, n_2, \ldots, n_r$ be positive integers whose sum is n, that is, $n_1 + n_2 + \cdots + n_r = n$. Then there exist

$$\frac{n!}{n_1!n_2!n_3!\cdots n_r!}$$

different ordered partitions of A of the form $[A_1, A_2, \ldots, A_r]$ where A_1 contains n_1 elements, A_2 contains n_2 elements,..., and A_r contains n_r elements.

Solved Problem 4.1 Suppose a license plate contains two letters followed by three digits with the first digit not zero. How many different license plates can be printed?

Solution. Each letter can be printed in 26 different ways, the first digit in 9 ways and each of the other two digits in 10 ways. Hence

$$26 \cdot 26 \cdot 9 \cdot 10 \cdot 10 = 608{,}400$$

different plates can be printed.

Solved Problem 4.2 How many seven-letter permutations can be formed using the letters of the word "BENZENE"?

Solution. We seek the number of permutations of seven objects of which three are alike (the three Es), and two are alike (the two Ns). By Theorem 4.4, the number of such words is

$$P(7;3,2) = \frac{7!}{3!2!} = \frac{7 \cdot 6 \cdot 5 \cdot 4 \cdot 3 \cdot 2 \cdot 1}{3 \cdot 2 \cdot 1 \cdot 2 \cdot 1} = 420$$

Solved Problem 4.3 Find the number of m ways that 12 students can be partitioned into three teams, A_1, A_2, and A_3, so that each team contains four students.

Solution. Let A denote one of the students. Then there are $\binom{11}{3}$ ways to choose three other students to be on the same team as A. Now let B denote a student who is not on the same team as A; then there are $\binom{7}{3}$ ways to choose three students of the remaining students to be on the same team as B. The remaining four students constitute the third team. Thus, altogether, the number of ways to partition the student is

$$m = \binom{11}{3} \cdot \binom{7}{3} = 165 \cdot 35 = 5775$$

Chapter 5
RELATIONS

IN THIS CHAPTER:

✔ *Product Sets*
✔ *Relations*
✔ *Pictorial Representations of Relations*
✔ *Composition of Relations*
✔ *Other Types of Relations*
✔ *Closure Properties*
✔ *Equivalence Relations*
✔ *Partial Ordering Relations*

Product Sets

Consider two arbitrary sets A and B. The set of all ordered pairs (a, b) where $a \in A$ and $b \in B$ is called the *product*, or *Cartesian product*, of A and B. A short designation of this product is $A \times B$, which is read "A cross B." By definition,

$$A \times B = \{(a,b): a \in A \text{ and } b \in B\}$$

Example 5.1 Let $A = \{1, 2\}$ and $B = \{a, b, c\}$. Then

$$A \times B = \{(1,a),(1,b),(1,c),(2,a),(2,b),(2,c)\}$$
$$B \times A = \{(a,1),(a,2),(b,1),(b,2),(c,1),(c,2)\}$$

Also $\qquad\qquad A \times A = \{(1,1),(1,2),(2,1),(2,2)\}$

There are two things worth noting in the above example. First of all, $A \times B \neq B \times A$. The Cartesian products deal with ordered pairs, so naturally the order in which the sets are considered is important. Secondly, using $n(S)$ for the number of elements in a set S, we have

$$n(A \times B) = 6 = 2 \cdot 3 = n(A) \cdot n(B)$$

In fact, $n(A \times B) = n(A) \cdot n(B)$ for any finite sets A and B. This follows from the observation that, for an ordered pair (a, b) in $A \times B$, there are $n(A)$ possibilities for a, and for each of these, there are $n(B)$ possibilities for b.

The idea of a product of sets can be extended to any finite number of sets. For any sets $A_1, A_2,..., A_n$, the sets of all ordered n-tuples $(a_1, a_2,... a_n)$ where $a_1 \in A_1, a_2 \in A_2,..., a_n \in A_n$ is called the *product* of the sets $A_1,..., A_n$ and is denoted by

$$A_1 \times A_2 \times \cdots \times A_n \qquad \text{or} \qquad \prod_{i=1}^{n} A_i$$

Relations

We begin with a definition.

Definition. Let A and B be sets. A *binary relation* or, simply, *relation* from A to B is a subset of $A \times B$.

Suppose R is a relation from A to B. Then R is a set of ordered pairs where each first element comes from A and each second element comes from B. That is, for each pair $a \in A$ and $b \in B$, exactly one of the following is true:

(i) $(a, b) \in R$; we then say "a is R-related to b" and write aRb.

(ii) $(a, b) \notin R$; we then say "a is not R-related to b" and write $a\slashed{R}b$.

If R is a relation from a set A to itself, that is, if R is a subset of $A^2 = A \times A$, then we say that R is a relation on A.

The *domain* of a relation R is the set of all first elements of the ordered pairs that belong to R, and the *range* of R is the set of second elements.

Inverse Relation

Let R be any relation from a set A to a set B. The *inverse* of R, denoted by R^{-1}, is the relation from B to A which consists of those ordered pairs which, when reversed, belong to R; that is,

$$R^{-1} = \{(b,a):(a,b) \in R\}$$

For example, the inverse of the relation $R = \{(1, y), (1, z), (3, y)\}$ from $A = \{1, 2, 3\}$ to $B = \{x, y, z\}$ is given by:

$$R^{-1} = \{(y,1),(z,1),(y,3)\}$$

Clearly, if R is any relation, then $\left(R^{-1}\right)^{-1} = R$. Also, the domain and range of R^{-1} are equal, respectively, to the range and domain of R. Moreover, if R is a relation on A, then R^{-1} is also a relation on A.

Functions as Relations

There is another point of view from which functions may be considered. First of all, every function $f: A \to B$ *gives rise to a relation from A to B* called the *graph of f* and defined by

$$\text{Graph of } f = \{(a, b): a \in A, b = f(a)\}$$

Two functions $f: A \to B$ and $g: A \to B$ are defined to be equal, written $f = g$, if $f(a) = g(a)$ for every $a \in A$; that is, if they have the same graph. Now, such a graph relation has the property that each a in A belongs to a unique ordered pair (a, b) in the relation. On the other hand, any relation f from A to B that has this property gives rise to a function $f: A \to B$, where $f(a)$

$= b$ for each (a,b) in f. Consequently, one may equivalently define a function as follows:

Definition: A function $f: A \rightarrow B$ is a relation from A to B (i.e., a subset of $A \times B$) such that each $a \in A$ belongs to a unique ordered pair (a,b) in f.

Pictorial Representations of Relations

First, we consider a relation S on the set $\mathbf{R}$ of real numbers; that is, S is a subset of $\mathbf{R}^2 = \mathbf{R} \times \mathbf{R}$. Since $\mathbf{R}^2$ can be represented by the set of points in the plane, we can picture S by emphasizing those points in the plane that belong to S. The pictorial representation of the relation is sometimes called the *graph* of the relation.

Frequently, the relation S consists of all ordered pairs of real numbers which satisfy some given equation

$$E(x, y) = 0$$

In this case, the graph of the relation is the same as the graph of the equation.

Representations of Relations on Finite Sets

Suppose A and B are finite sets. The following are two ways of picturing a relation R from A to B.

(i) From a rectangular array whose rows are labeled by the elements of A and whose columns are labeled by the elements of B. Put a 1 or 0 in each position of the array according to $a \in A$ is or is not related to $b \in B$. This array is called the *matrix of the relation*.

(ii) Write down the elements of A and the elements of B in two disjoint disks, and then draw an arrow from $a \in A$ to $b \in B$ whenever a is related to b. This picture will be called the *arrow diagram* of the relation.

Directed Graphs of Relations on Sets

There is another way of picturing a relation R when R is a relation from a finite set to itself. First we write down the elements of the set, and then we draw an arrow from each element x to each element y whenever x is related to y. This diagram is called the *directed graph* of the relation. Figure 5-1, for example, shows the directed graph of the following relation R on the set $A = \{1, 2, 3, 4\}$:

$$R = \{(1, 2), (2, 2), (2, 4), (3, 2), (3, 4), (4, 1), (4, 3)\}$$

Observe that there is an arrow from 2 to itself, since 2 is related to 2 under R.

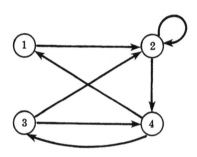

Figure 5-1

Composition of Relations

Let A, B, and C be sets, and let R be a relation from A to B and let S be a relation from B to C. That is, R is a subset of $A \times B$ and S is a subset of $B \times C$. Then R and S give rise to a relation from A to C denoted by $R \circ S$ and defined by

$$a(R \circ S)c \text{ if for some } b \in B, \text{ we have } aRb \text{ and } bSc$$

That is,

$$R \circ S = \{(a,c) : \text{there exists } b \in B \text{ for which } (a,b) \in R \text{ and } (b,c) \in S\}$$

The relation $R \circ S$ is called the *composition* of R and S; it is sometimes denoted simply by RS.

Suppose R is a relation on a set A, that is, R is a relation from a set A to itself. Then $R \circ R$, the composition of R with itself is always defined, and $R \circ R$ is sometimes denoted by R^2. Similarly, $R^3 = R^2 \circ R = R \circ R \circ R \circ$ and so on. Thus, R^n is defined for all positive n.

Other Types of Relations

Consider a given set A. This section discusses a number of important types of relations that are defined on A.

Reflexive Relations

A relation R on a set A is *reflexive* if aRa for every $a \in A$, that is, if $(a, a) \in R$ for every $a \in A$. Thus, R is not reflexive if there exists an $a \in A$ such that $(a, a) \notin R$.

Symmetric and Antisymmetric Relations

A relation R on a set A is *symmetric* if whenever aRb then bRa, that is, if whenever $(a, b) \in R$ then $(b, a) \in R$. Thus, R is not symmetric if there exists $a, b \in A$ such that $(a, b) \in R$ but $(b, a) \notin R$.

A relation R on a set A is *antisymmetric* if whenever aRb and bRa, then $a = b$, that is, if whenever $(a, b), (b, a) \in R$ then $a = b$. Thus R is not antisymmetric if there exist $a, b \in A$ such that (a, b) and (b, a) belong to R, but $a \neq b$.

Transitive Relations

A relation R on a set A is *transitive* if whenever aRb and bRc then aRc, that is, if whenever $(a, b), (b, c) \in R$ then $(a, c) \in R$. Thus, R is not transitive if there exist $a, b, c \in A$ such that $(a, b), (b, c) \in R$ but $(a, c) \notin R$.

The property of transitivity can also be expressed in terms of the composition of relations. For a relation R on A, we define

$$R^2 = R \circ R \text{ and, more generally,} \qquad R^n = R^{n-1} \circ R$$

Then we have the following result.

Theorem 5.1: A relation R is transitive if and only if $R^n \subseteq R$ for $n \geq 1$.

Closure Properties

Consider a given set A and the collection of all relations on A. Let P be a property of such relations, such as being symmetric or being transitive. A relation with property P will be called a P-relation. The P-*closure* of an arbitrary relation R on A, written P(R), is a P-relation such that

$$R \subseteq P(R) \subseteq S$$

for every P-relation S containing R. We will write
$$\text{reflexive}(R), \qquad \text{symmetric}(R), \qquad \text{and transitive}(R)$$
for the reflexive, symmetric, and transitive closures of R.

Reflexive and Symmetric Closures

The next theorem tells us how to easily obtain the reflexive and symmetric closures of a relation. Here $\Delta_A = \{(a,a): a \in A\}$ is the *diagonal* or *equality* relation on A.

Theorem 5.2: Let R be a relation on a set A. Then:

(i) $R \cup \Delta_A$ is the reflexive closure of R.
(ii) $R \cup R^{-1}$ is the symmetric closure of R.

In other words, reflexive(R) is obtained by simply adding to R those elements (a, a) in the diagonal which do not already belong to R, and symmetric(R) is obtained by adding to R all pairs (b, a) whenever (a, b) belongs to R.

Transitive Closure

Let R be a relation on a set A. Recall that $R^2 = R \circ R$ and $R^n = R^{n-1} \circ R$. We define

$$R^* = \bigcup_{i=1}^{\infty} R^i$$

The following theorem applies.

Theorem 5.3: R^* is the transitive closure of a relation R.

Suppose A is a finite set with n elements. Then:

$$R^* = R \cup R^2 \cup \cdots \cup R^n$$

This gives us the following result.

Theorem 5.4: Let R be a relation on a set A with n elements. Then:

$$\text{transitive}(R) = R \cup R^2 \cup \cdots \cup R^n$$

Equivalence Relations

Consider a nonempty set S. A relation R on S is an *equivalence relation* if R is reflexive, symmetric, and transitive. That is, R is an equivalence relation on S if it has the following three properties:

1. For every $a \in S$, aRa.
2. If aRb, then bRa.
3. If aRb and bRc, then aRc.

The general idea behind an equivalence relation is that it is a classification of objects that are in some way "alike." In fact, the relation "=" of equality on any set S is an equivalence relation; that is:

1. $a = a$ for every $a \in S$.
2. If $a = b$, then $b = a$.
3. If $a = b$ and $b = c$, then $a = c$.

Equivalence Relations and Partitions

This subsection explores the relationship between equivalence relations and partitions on a nonempty set S. Recall first that a partition P of S is a collection $\{A_i\}$ of nonempty subsets of S with the following two properties:

(1) Each $a \in S$ belongs to some A_i.
(2) If $A_i \neq A_j$, then $A_i \cap A_j = \varnothing$.

In other words, a partition P of S is a subdivision of S into disjoint nonempty sets.

Suppose R is an equivalence relation on a set S. For each a in S, let $[a]$ denote the set of elements of S to which a is related under R; that is,

$$[a] = \{x : (a, x) \in R\}$$

We call $[a]$ the *equivalence class* of a in S; any $b \in [a]$ is called a *representative* of the equivalence class.

The collection of all equivalence classes of elements of S under an equivalence relation R is denoted by S/R, that is,

$$S/R = \{[a] : a \in S\}$$

It is called the *quotient* set of S by R. The fundamental property of a quotient set is contained in the following theorem.

Theorem 5.5: Let R be an equivalence relation on a set S. Then the quotient set S/R is a partition of S. Specifically:

(i) For each a in S, we have $a \in [a]$.
(ii) $[a] = [b]$ if and only if $(a, b) \in R$.
(iii) If $[a] \neq [b]$, then $[a]$ and $[b]$ are disjoint.

Conversely, given a partition $\{A_i\}$ of the set S, there is an equivalence relation R on S such that the sets A_i are the equivalence classes.

Partial Ordering Relations

This section defines another important class of relations. A relation R on a set is called a *partial ordering* or a *partial order* if R is reflexive, anti-symmetric, and transitive. A set S together with a partial ordering R is called a *partially ordered set* or *poset*. For example, $\leq$ is a partial order on the set of real numbers, and set containment is a partial order on the powerset of a set.

Solved Problem 5.1 Given $A = \{1, 2\}$, $B = \{x, y, z\}$, and $C = \{3, 4\}$. Find: $A \times B \times C$.

Solution. $A \times B \times C$ consists of all ordered triplets (a, b, c) where $a \in A$, $b \in B$, $c \in C$. $A \times B \times C$ can be systematically obtained by a so-called tree diagram (Figure 5-2). The elements of $A \times B \times C$ are precisely the 12 ordered triplets to the right of the tree diagram.

Observe that $n(A) = 2$, $n(B) = 3$, and $n(C) = 2$ and, as expected,

$$n(A \times B \times C) = 12 = n(A) \cdot n(B) \cdot n(C)$$

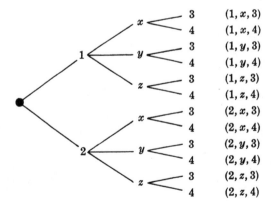

Figure 5-2

Solved Problem 5.2 Find the number of relations from $A = \{a, b, c\}$ to $B = \{1, 2\}$.

Solution. There are $3(2) = 6$ elements in $A \times B$, and hence there are $m = 2^6 = 64$ subsets of $A \times B$. Thus, there are $m = 64$ relations from A to B.

Introduction; Data Structures

Graphs, directed graphs, and trees and binary trees appear in many areas of mathematics and computer science. However, in order to understand how these objects may be stored in memory and to understand algorithms on them, we need to know a little about certain data structures. We will now discuss linked lists and pointers, and stacks and queues.

Linked Lists and Pointers

Linked lists and pointers will be introduced by means of an example. Suppose a brokerage firm maintains a file in which each record contains a customer's name and salesman; say the file contains the following data:

Customer	Adams	Brown	Clark	Drew	Evans	Farmer	Geller	Hill	Infeld
Salesman	Smith	Ray	Ray	Jones	Smith	Jones	Ray	Smith	Ray

There are two basic operations that one would want to perform on the data:

Operation A: Given the name of a customer, find his salesman.

Operation B: Given the name of a salesman, find the list of his customers.

We discuss a number of ways the data may be stored in the computer, and the ease with which one can perform the operations *A* and *B* on the data.

Clearly, the file could be stored in the computer by an array with two rows (or columns) of nine names. Since the customers are listed alphabetically, one could easily perform Operation *A*. However, in order to perform Operation *B*, one must search through the entire array.

One can easily store the data in memory using a two-dimensional array where, say, the rows correspond to an alphabetical listing of the salesman, and where there is a 1 in the matrix indicating the salesman of a customer and there are 0s elsewhere. The main drawback of such a representation is that there may be a waste of a lot of memory because many 0s may be in the matrix. For example, if a firm has 1000 customers and 20 salesmen, one would need 20,000 memory locations for the data, but only 1000 of them would be useful.

We discuss below a way of storing the data in memory which uses linked lists and pointers. By a *linked list*, we mean a linear collection of data elements, called *nodes*, where the linear order is given by means of

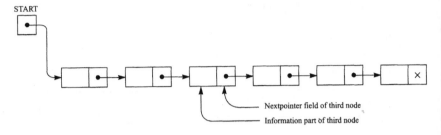

Figure 6-1

a field of pointers. Figure 6-1 is a schematic diagram of a linked list with six nodes.

That is, each node is divided into two parts: the first part contains the information of the element (e.g., NAME, ADDRESS,...), and the second part, called the *linked field* or *nextpointer field*, contains the address of the next node in the list. This pointer field is indicated by an arrow drawn from one node to the next node in the list. There is also a variable pointer, called START in Figure 6-1, which gives the address of the first node in the list. Furthermore, the pointer field of the last node contains an invalid address, called a *null pointer*, which indicates the end of the list.

One main way of storing the original data, pictured in Figure 6-2, uses linked lists. Observe that there are separate (sorted alphabetically) arrays for the customers and the salesmen.

	Customer	SLSM	NEXT
1	Adams	3	5
2	Brown	2	3
3	Clark	2	7
4	Drew	1	6
5	Evans	3	8
6	Farmer	1	0
7	Geller	2	9
8	Hill	3	0
9	Infeld	2	0

	Salesman	START	
	Jones	4	1
	Ray	2	2
	Smith	1	4

Figure 6-2

Also, there is a pointer array SLSM parallel to CUSTOMER which gives the location of the salesman of a customer; hence operation A can be performed very easily and quickly. Furthermore, the list of customers of each salesman is a linked list as discussed above. Specifically, there is a pointer array START parallel to SALESMAN which points to the first customer of a salesman, and there is an array NEXT that points to the location of the next customer in the salesman's list (or contains a 0 to indicate the end of the list). This process is indicated by the arrows in Figure 6-2 for the salesman Ray.

Operation B can now be performed easily and quickly; that is, one does not need to search through the list of all customers in order to obtain the list of customers of a given salesman. The following is such an algorithm (which is written in pseudocode).

Algorithm 6.1   The name of a salesman is read and the list of his customers is printed.

Step 1.   Read XXX.

Step 2.   Find K such that SALESMAN[K] = XXX. [Use binary search.]

Step 3.   Set PTR := START[K]. Initializes pointer PTR.]

Step 4.   Repeat while PTR $\neq$ NULL.
(*a*)   Print CUSTOMER[PTR].
(*b*)   Set PTR := NEXT[PTR].   [Update PTR.]
{End of loop.}

Step 5.   Exit.

Stacks, Queues, and Priority Queues

There are data structures other than arrays and linked lists that will occur in our graph algorithms. These structures, stacks, queues, and priority queues, are briefly described below.

(*a*) **Stack**: A *stack*, also called a *last-in-first-out* (LIFO) system, is a linear list in which insertions and deletions can take place only at one end, called the "top" of the list. This structure is similar in its operation to a

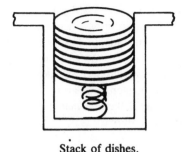

Stack of dishes.

Figure 6-3

stack of dishes on a spring system, as pictured in Figure 6-3. Note that new dishes are inserted only at the top of the stack and dishes can be deleted only from the top of the stack.

(*b*) **Queue**: A *queue*, also called a *first-in-first-out* (FIFO) system, is a linear list in which deletions can only take place at one end of the list, the "front" of the list, and insertions can only take place at the other end of the list, the "rear" of the list. The structure operates in much the same way as a line of people waiting at a bus stop, as pictured in Figure 6-4. That is, the first person in line is the first person to board the bus, and a new person goes to the end of the line.

Queue waiting for a bus.

Figure 6-4

(*c*) *Priority queue*: Let *S* be a set of elements where new elements may be periodically inserted, but where the current largest element (element with the "highest priority") is always deleted. Then *S* is called a *priority queue*. The rules "women and children first" and "age before beauty" are examples of priority queues. Stacks and ordinary queues are special kinds of priority queues. Specifically, the element with the highest priority in a stack is the last element inserted, but the element with the highest priority in a queue is the first element inserted.

Graphs and Multigraphs

A *graph G* consists of two things:

    (i) A set $V = V(G)$ whose elements are called *vertices, points*, or *nodes* of *G*.

    (ii) A set $E = E(G)$ of unordered pairs of distinct vertices called *edges* of *G*.

We denote such a graph by $G(V, E)$ when we want to emphasize the two parts of *G*.

Vertices *u* and *v* are said to be *adjacent* if there is an edge $e = \{u, v\}$. In such a case, *u* and *v* are called the *endpoints* of *e*, and *e* is said to *connect u* and *v*. Also, the edge *e* is said to be *incident* on each of its endpoints *u* and *v*.

Graphs are pictured by diagrams in the plane in a natural way. Specifically, each vertex *v* in *V* is represented by a dot (or small circle), and each edge $e = \{v_1, v_2\}$ is represented by a curve which connects its endpoints v_1 and v_2.

Multigraphs

Consider the diagram in Figure 6-5. The edges e_4 and e_5 are called *multiple edges* since they connect the same endpoints, and the edge e_6 is called a *loop* since its endpoints are the same vertex. Such a diagram is called a *multigraph*; the formal definition of a graph permits neither multiple edges nor loops. Thus, a graph may be defined to be a multigraph without multiple edges or loops.

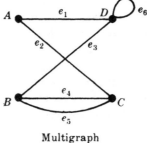

Multigraph

Figure 6-5

Degree of a Vertex

The *degree* of a vertex v in a graph G, written deg (v), is equal to the number of edges in G which contain v, that is, which are incident on v. Since each edge is counted twice in counting the degrees of the vertices of G, we have the following simple but important result.

Theorem 6.1: The sum of the degrees of the vertices of a graph G is equal to twice the number of edges in G.

A vertex of degree zero is called an *isolated* vertex.

Finite Graphs; Trivial Graph

A multigraph is said to be *finite* if it has a finite number of vertices and a finite number of edges. Observe that a graph with a finite number of vertices must automatically have a finite number of edges and so must be finite. The finite graph with one vertex and no edges, i.e., a single point, is called the *trivial graph*.

Subgraphs; Isomorphic and Homeomorphic Graphs

Subgraphs

Consider a graph $G = G(V, E)$. A graph $H = H(V', E')$ is called a *subgraph* of G if the vertices and edges of H are contained in the vertices and edges of G, that is, if $V' \subseteq V$ and $E' \subseteq E$. In particular:

(i) A subgraph $H(V', E')$ of $G(V, E)$ is called the subgraph *induced* by its vertices V' if its edge set E' contains all edges in G whose end-points belong to vertices in H.

(ii) If v is a vertex in G, the $G - v$ is the subgraph of G obtained by deleting v from G and deleting all edges in G which contain v.

(iii) If e is an edge in G, then $G - e$ is the subgraph of G obtained by simply deleting the edge e from G.

Isomorphic Graphs

Graphs $G = G(V, E)$ and $G^* = H(V^*, E^*)$ are said to be isomorphic if there exists a one-to-one correspondence $f: V \rightarrow V^*$ such that $\{u, v\}$ is an edge of G if and only if $\{f(u), f(v)\}$ is an edge of G^*. Normally, we do not distinguish between isomorphic graphs (even though their diagrams may "look different"). Figure 6-6 gives ten graphs pictured as letters. We note that A and R are isomorphic graphs. Also, F and T, K and X, and M, S, V, and Z are isomorphic graphs.

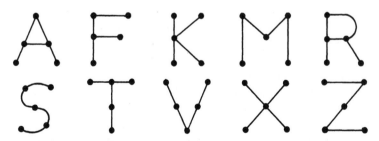

Figure 6-6

Homeomorphic Graphs

Given any graph G, we can obtain a new graph by dividing an edge of G with additional vertices. Two graphs G and G^* are said to be *homeomorphic* if they can be obtained from the same graph or isomorphic graphs by this method. The graphs (a) and (b) in Figure 6-7 are not isomorphic, but they are homeomorphic since they can be obtained from the graph (c) by adding appropriate vertices.

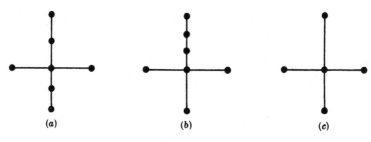

(a) $\qquad$ (b) $\qquad$ (c)

Figure 6-7

Paths; Connectivity

A *path* in a multigraph G consists of an alternating sequence of vertices and edges of the form

$$v_0, e_1, v_1, e_2, v_2, \ldots, e_{n-1}, v_{n-1}, e_n, v_n$$

where each edge e_i contains the vertices v_{i-1} and v_i (which appear on the sides of e_i in the sequence). The number n of edges is called the *length* of the path. When there is no ambiguity, we denote a path by its sequence of vertices $(v_0, v_1, \ldots, v_n)$. The path is said to be *closed* if $v_0 = v_n$. Otherwise, we say the path is from v_0 to v_n, or *between* v_0 and v_n, or *connects* v_0 to v_n.

A *simple path* is a path in which all vertices are distinct. A path in which all edges are distinct will be called a *trail*. A *cycle* is a closed path of length 3 or more in which all vertices are distinct except $v_0 = v_n$. A cycle of length k is called a *k-cycle*.

By eliminating unnecessary edges, it is not difficult to see that any path from a vertex u to a vertex v can be replaced by a simple path from u to v. We state this result formally.

Theorem 6.2: There is a path from a vertex u to a vertex v if and only if there exists a simple path from u to v.

Connectivity; Connected Components

A graph G, is *connected* if there is a path between any two of its vertices. The graph in Figure 6-8(a) is connected, but the graph in Figure 6-8(b) is not connected since, for example, there is no path between vertices D and E.

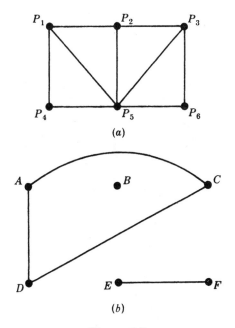

(a)

(b)

Figure 6-8

Suppose G is a graph. A connected subgraph H of G is called a *connected component* of G if H is not contained in any larger connected subgraph of G. It is intuitively clear that any graph G can be partitioned into its connected components. For example, the graph G in Figure 6-8(b) has three connected components, the subgraphs induced by the vertex sets $\{A, C, D\}$, $\{E, F\}$, and $\{B\}$.

The vertex B in Figure 6-8(b) is called an *isolated vertex* since B does not belong to any edge or, in other words, deg $(B) = 0$. Therefore, as noted, B itself forms a connected component of the graph.

Distance and Diameter

Consider a connected graph G. The *distance* between vertices u and v in G, written $d(u, v)$, is the length of the shortest path between u and v. The *diameter* of G, written diam(G), is the maximum distance between any two points in G. For example, in Figure 6-9(a), $d(A, F) = 2$ and diam(G) $= 3$ whereas in Figure 6-9(b), $d(A, F) = 3$ and diam(G) $= 4$.

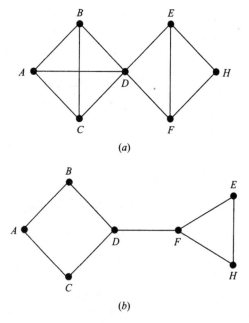

(a)

(b)

Figure 6-9

Cutpoints and Bridges

Let G be a connected graph. A vertex v in G is called a *cutpoint* if $G - v$ is disconnected. An edge e of G is called a *bridge* if $G - e$ is disconnected. In Figure 6-9(a), the vertex D is a cutpoint and there are no bridges. In Figure 6-9(b), the edge $e = \{D, F\}$ is a bridge.

Labeled and Weighted Graphs

A graph G is called a *labeled graph* if its edges and/or vertices are assigned data of one kind or another. In particular, G is called a *weighted graph* if each edge e of G is assigned a nonnegative number $w(e)$ called the *weight* or *length* of v. Figure 6-10 shows a weighted graph where the weight of each edge is given in the obvious way. The *weight* (or *length*) of a path in such a weighted graph G is defined to be the sum of the weights of the edges in the path. One important problem in graph theory is to find a *shortest path*, that is, a path of minimum weight (length), between any two given vertices. The length of a shortest path between P and Q in Figure 6-10 is 14.

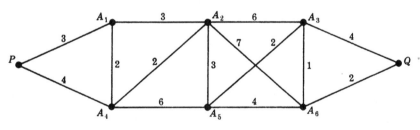

Figure 6-10

Complete, Regular, and Bipartite Graphs

There are many different types of graphs. This section considers three of them: complete, regular, and bipartite graphs.

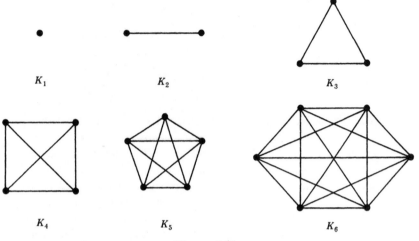

Figure 6-11

Complete Graphs

A graph G is said to be *complete* if every vertex in G is connected to every other vertex in G. Thus a complete graph G must be connected. The complete graph with n vertices is denoted by K_n. Figure 6-11 shows the graphs K_1 through K_6.

Regular Graphs

A graph G is *regular of degree k* or *k-regular* if every vertex has degree k. In other words, a graph is regular if every vertex has the same degree.

The connected regular graphs of degrees 0, 1, or 2 are easily described. The connected 0-regular graph is the trivial graph with one vertex and no edges. The connected 1-regular graph is the graph with two vertices and one edge connecting them. The connected 2-regular graph with n vertices is the graph which consists of a single n-cycle. See Figure 6-12.

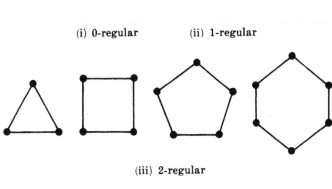

(i) 0-regular   (ii) 1-regular

(iii) 2-regular

Figure 6-12

The 3-regular graphs must have an even number of vertices since the sum of the degrees of the vertices is an even number (Theorem 6.1). Figure 6-13 shows two connected 3-regular graphs with six vertices. In general, regular graphs can be quite complicated. For example, there are nineteen 3-regular graphs with ten vertices. We note that the complete graph with n vertices K_n is regular of degree $n - 1$.

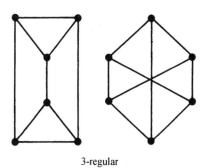

3-regular

Figure 6-13

Bipartite Graphs

A graph G is said to be *bipartite* if its vertices V can be partitioned into two subsets M and N such that each edge of G connects a vertex of M to a vertex of N. By a complete bipartite graph, we mean that each vertex of M is connected to each vertex of N; this graph is denoted by $K_{m,n}$ where m is the number of vertices in M and n is the number of vertices in N, and, for standardization, we will assume $m \leq n$. Figure 6-14 shows the graphs $K_{2,3}$, $K_{3,3}$, and $K_{2,4}$. Clearly, the graph $K_{m,n}$ has mn edges.

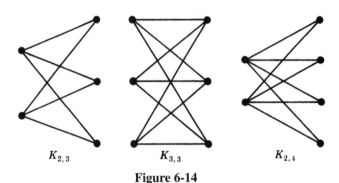

$K_{2,3}$ $K_{3,3}$ $K_{2,4}$

Figure 6-14

Tree Graphs

A graph T is called a *tree* if T is connected and T has no cycles. Examples of trees are shown in Figure 6-15. A *forest G* is a graph with no cycles; hence the connected components of a forest G are trees. The tree consisting of a single vertex with no edges is called the *degenerate tree*.

Consider a tree T. Clearly, there is only one simple path between two vertices of T; otherwise, the two paths would form a cycle. Also:

(*a*) Suppose there is no edge $\{u, v\}$ in T and we add the edge $e = \{u, v\}$ to T. Then the simple path from u to v in T and e will form a cycle; hence T is no longer a tree.

(*b*) On the other hand, suppose there is an edge $e = \{u, v\}$ in T, and we delete e from T. Then T is no longer connected (since there cannot be a path from u to v); hence T is no longer a tree.

The following theorem applies when our graphs are finite:

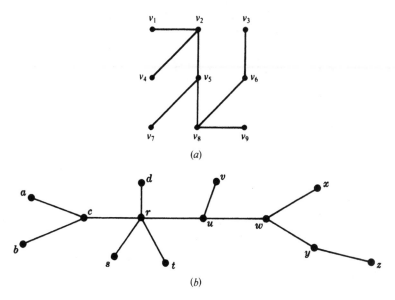

(a)

(b)

Figure 6-15

Theorem 6.3: Let G be a graph with $n > 1$ vertices. Then the following are equivalent:

   (i) G is a tree.
  (ii) G is cycle-free and has $n - 1$ edges.
 (iii) G is connected and has $n - 1$ edges.

This theorem also tells us that a finite tree T with n vertices must have $n - 1$ edges. For example, the tree in Figure 6-15(a) has 9 vertices and 8 edges, and the tree in Figure 6-15(b) has 13 vertices and 12 edges.

Spanning Trees

A subgraph T of a connected graph G is called a *spanning tree* of G if T is a tree and T includes all the vertices of G. Figure 6-16 shows a connected graph G and spanning trees T_1, T_2, and T_3 of G.

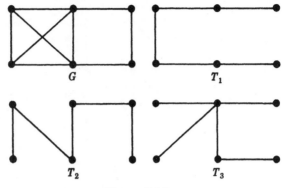

Figure 6-16

Minimum Spanning Trees

Suppose G is a connected weighted graph. That is, each edge of G is assigned a nonnegative number called the *weight* of the edge. Then any spanning tree T of G is assigned a total weight obtained by adding the weights of the edges in T. A *minimal spanning tree* of G is a spanning tree whose total weight is as small as possible.

Algorithms 6.2A and 6.2B, which follow, enable us to find a minimal spanning tree T of a connected weighted graph G where G has n vertices. (In which case, T must have $n - 1$ edges.)

Algorithm 6.2A   The input is a connected weighted graph G with n vertices.

Step 1.   Arrange the edges of G in the order of decreasing weights.

Step 2.   Proceeding sequentially, delete each edge that does not disconnect the graph until $n - 1$ edges remain.

Step 3.   Exit.

Algorithm 6.2B (Kruskal)   The input is a connected weighted graph G with n vertices.

Step 1.   Arrange the edges of G in the order of increasing weights.

Step 2.   Starting only with the vertices of G and proceedimg sequentially, add each edge that does not result in a cycle until $n - 1$ edges are added.

Step 3.   Exit.

The weight of a minimal spanning tree is unique, but the minimal spanning tree itself is not. Different minimal spanning trees can occur when two or more edges have the same weight. In such a case, the arrangement of the edges in Step 1 of Algorithms 6.2A or 6.2B is not unique and hence may result in different minimal spanning trees.

Solved Problem 6.1 Consider the graph G in Figure 6-17. Find the subgraphs obtained when each vertex is deleted. Does G have any cut points?

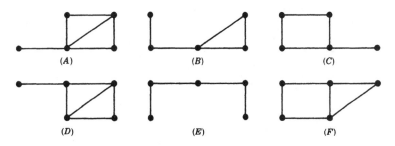

Figure 6-17

Solution. When we delete a vertex from G, we also have to delete all edges which contain the vertex. The six graphs obtained by deleting each of the vertices of G are shown in Figure 6-17. All six graphs are connected; hence no vertex is a cut point.

Solved Problem 6.2 Find all spanning trees of the graph G shown in Figure 6-18(a).

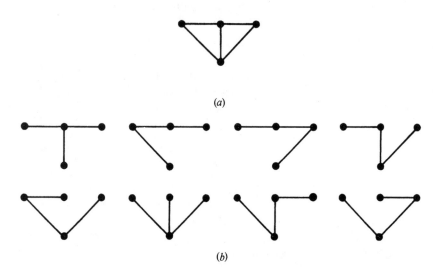

(a)

(b)

Figure 6-18

Solution. There are eight such spanning trees as shown in Figure 6-18(b). Each spanning tree must have $4 - 1 = 3$ edges since G has four vertices. Thus, each tree can be obtained by deleting two of the five edges of G. This can be done in 10 ways, except that two of the ways lead to disconnected graphs. Hence the above eight spanning trees are all the spanning trees of G.

Chapter 7
BINARY TREES

IN THIS CHAPTER:

- ✔ *Binary Trees*
- ✔ *Complete and Extended Binary Trees*
- ✔ *Representing Binary Trees in Memory*
- ✔ *Traversing Binary Trees*
- ✔ *Binary Search Tree*

Binary Trees

A *binary tree T* is defined as a finite set of elements, called *nodes* such that:

1. *T* is empty (called the *null tree* or *empty tree*), or
2. *T* contains a distinguished node *R*, called the *root* of *T*, and the remaining nodes of *T* form an ordered pair of disjoint binary trees T_1 and T_2.

If *T* does contain a root *R*, then the two trees T_1 and T_2 are called, respectively, the *left* and *right subtrees* of *R*. If T_1 is nonempty, then its root is called the *left successor* of R; similarly, if T_2 is nonempty, then its root is called the *right successor* of R.

The above definition of a binary tree T is recursive since T is defined in terms of the binary subtrees T_1 and T_2. This means, in particular, that every node N in T has 0, 1, or 2 successors. A node with no successors is called a *terminal node*. Thus both subtrees of a terminal node are empty.

Picture of a Binary Tree

A binary tree T is frequently presented by a diagram in the plane called a *picture* of T. Specifically, the diagram in Figure 7-1 represents a binary tree as follows:

(i) T consists of 11 nodes, represented by the letters A through L, excluding I.
(ii) The root of T is the node A at the top of the diagram.
(iii) A left-downward slanted line at a node N indicates a left successor of N; and a right-downward slanted line at N indicates a right successor of N.

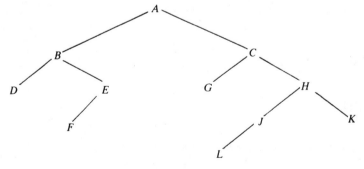

Figure 7-1

Accordingly, in Figure 7-1:

(*a*) B is a left successor and C is a right successor of the root A.

(b) The left subtree of the root A consists of the nodes B, D, E, and F, and the right subtree of A consists of the nodes C, G, H, J, K, and L.

(c) The nodes A, B, C, and H have two successors, the nodes E and J have only one successor, and the nodes D, F, G, L, and K have no successors, i.e., they are terminal nodes.

Algebraic Expressions

Consider any algebraic expression E involving only binary operations, such as

$$E = (a - b) / ((c * d) + e)$$

E can be represented by means of the binary tree T pictured in Figure 7-2. That is, each variable or constant in E appears as an "internal" node in T whose left and right subtrees correspond to the operands of the operation. For example:

(a) In the expression E, the operands of $+$ are $c * d$ and e.

(b) In the tree T, the subtrees of the node $+$ correspond to the subexpressions $c * d$ and e.

Clearly, every algebraic expression will correspond to a unique tree, and vice versa.

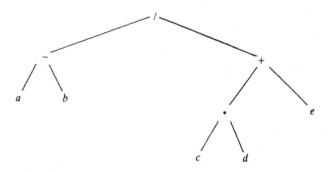

$$E = (a - b) / ((c * d) + e).$$

Figure 7-2

Terminology

Terminology describing family relationships is frequently used to describe relationships between the nodes of a tree T. Specifically, suppose N is a node in T with left successor S_1 and right successor S_2. Then N is called the *parent* (or *father*) of S_1 and S_2. Analogously, S_1 is called the *left child* (or *son*) of N, and S_2 is called the *right child* (or *son*) of N. Furthermore, S_1 and S_2 are said to be *siblings* (or *brothers*). Every node N in a binary tree T, except the root, has a unique parent, called the *predecessor* of N.

The terms descendant and ancestor have their usual meaning. That is, a node L is called a *descendant* of a node N (and N is called an *ancestor* of L) if there is a succession of children from N to L. In particular, L is called a *left* or *right descendant* of N according to whether L belongs to the left or right subtree of N.

Terminology from graph theory and horticulture is also used with a binary tree T. Specifically, the line drawn from a node N of T to a successor is called an *edge*, and a sequence of consecutive edges is called a *path*. A terminal node is called a *leaf*, and a path ending in a leaf is called a *branch*.

Each node in a binary tree T is assigned a *level number*, as follows. The root R of the tree T is assigned the level number 0, and every other node is assigned a level number which is 1 more than the level number of its parent. Furthermore, those nodes with the same level number are said to belong to the same *generation*.

The *depth* (or *height*) of a tree T is the maximum number of nodes in a branch of T. This turns out to be 1 or more than the largest level number of T. The tree T in Figure 7-1 has depth 5.

Complete and Extended Binary Trees

This section considers two special kinds of binary trees.

Complete Binary Trees

Consider any binary tree T. Each node of T can have at most two children. Accordingly, one can show that level r of T can have at most 2^r

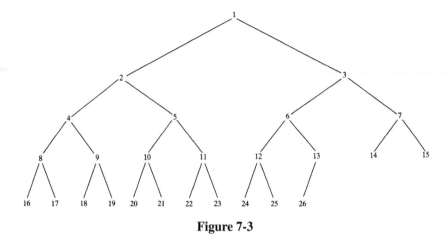

Figure 7-3

nodes. The tree T is said to be *complete* if all its levels, except possibly the last, have the maximum number of possible nodes, and if all the nodes at the last level appear as far left as possible. Thus there is a unique complete tree T_n with exactly n nodes (we are, of course, ignoring the contents of nodes). The complete tree T_{26} with 26 nodes appears in Figure 7-3.

The nodes of the complete binary tree T_{26} in Figure 7-3 have been purposely labeled by the integers 1, 2,..., 26, from left to right, generation by generation. With this labeling, one can easily determine the children and parent of any node K in any complete tree T_n. Specifically, the left and right children of the node K are, respectively, $2 * K$ and $2 * K + 1$, and the parent of K is the node $[K/2]$. For example, the children of node 9 are the nodes 18 and 19, and its parent is the node $[9/2] = 4$. The depth d_n of the complete tree T_n with n nodes is given by

$$d_n = \lfloor \log_2 n + 1 \rfloor$$

This is a relatively small number. For example, if the complete tree T_n has $n = 1\,000\,000$ nodes, then its depth is $d_n = 21$.

Extended Binary Trees: 2-Trees

A binary tree T is said to be a *2-tree* or an *extended binary tree* if each node N has either 0 or 2 children. In such a case, the nodes with two children are called *internal nodes*, and the nodes with 0 children are called *external nodes*. Sometimes the nodes are distinguished in diagrams by using circles for internal nodes and squares for external nodes.

The term "extended binary tree" comes from the following operation. Consider any binary tree T, such as the tree in Figure 7-4(a). Then T may be "converted" into a 2-tree by replacing each empty subtree by a new node, as pictured in Figure 7-4(b). Observe that the new tree is, indeed, a 2-tree. Furthermore, the nodes in the original tree T are now the internal nodes in the extended tree, and the new nodes are the external nodes in the extended tree.

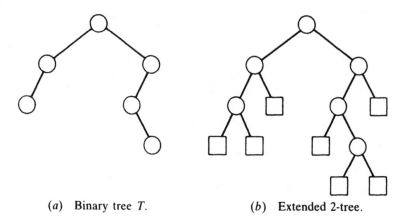

(a)   Binary tree T.   (b)   Extended 2-tree.

Figure 7-4

Representing Binary Trees in Memory

Let T be a binary tree. This section discusses two ways of representing T in memory. The first and usual way is called the link representation of T and is analogous to the way linked lists are represented in memory. The second way, which uses a single array, is called the sequential representation of T. The main requirement of any representation of T is that one should have direct access to the root R of T and, given any node N of T, one should have direct access to the children of N.

Linked Representation of Binary Trees

Consider a binary tree T. Unless other wise stated or implied, T will be maintained in memory by means of a *linked representation* that uses three parallel arrays, INFO, LEFT, and RIGHT, and a pointer variable ROOT as follows. First of all, each node N of T will correspond to a location K such that:

1. INFO[K] contains the data at the node N.
2. LEFT[K] contains the location of the left child of node N.
3. RIGHT[K] contains the location of the right child of node N.

Furthermore, ROOT will contain the location of the root R of T. If any subtree is empty, then the corresponding pointer will contain the null value; if the tree T itself is empty, then ROOT will contain the null value.

Sequential Representation of Binary Trees

Suppose T is a binary tree that is complete or nearly complete. Then there is an efficient way of maintaining T in memory called the *sequential representation* of T. This representation uses only a single linear array TREE together with a pointer variable END as follows:

(a) The root R of T is stored in TREE[1].
(b) If a node N occupies TREE[K], then its left child is stored in TREE[$2 * K$] and its right child is stored in TREE[$2 * K + 1$].
(c) END contains the location of the last node of T.

Furthermore, the node N at TREE[K] contains an empty left or right subtree according as $2 * K$ or $2 * K + 1$ exceeds END or according as TREE[$2 * K$] or TREE[$2 * K + 1$] contains the NULL value.

Traversing Binary Trees

There are three standard ways of traversing a binary tree T with root R. These three algorithms, called *preorder*, *inorder*, and *postorder*, are as follows:

Preorder:	(1) Process the root R.
	(2) Traverse the left subtree of R in preorder.
	(3) Traverse the right subtree of R in preorder.
Inorder:	(1) Traverse the left subtree of R in inorder.
	(2) Process the root R.
	(3) Traverse the right subtree of R in inorder.
Postorder:	(1) Traverse the left subtree of R in postorder.
	(2) Traverse the right subtree of R in postorder.
	(3) Process the root R.

Observe that each algorithm contains the same three steps, and that the left subtree of R is always traversed before the right subtree. The difference between the algorithms is the time at which the root R is processed. Specifically, in the "pre" algorithm, the root R is processed before the subtrees are traversed; in the "in" algorithm, the root R is processed between the traversals of the subtrees; and in the "post" algorithm, the root R is processed after the subtrees are traversed.

The three algorithms are sometimes called, respectively, the node-left-right (NLR) traversal, the left-node-right (LNR) traversal and the left-right-node (LRN) traversal.

Binary Search Trees

This section discusses one of the most important data structures in computer science, a binary search tree. This structure enables us to search for and find an element with an average running time $f(n) = O(\log_2 n)$, where n is the number of data items. It also enables us to easily insert and delete elements. This structure contrasts with the following structures:

(a) *Sorted linear array*: Here one can search for and find an element with running time $f(n) = O(\log_2 n)$. However, inserting and deleting elements is expensive since, on the average, it involves moving $O(n)$ elements.

(b) *Linked list*: Here one can easily insert and delete elements. However, it is expensive to search and find an element, since one must use a linear search with running time $f(n) = O(n)$.

Although each node in a binary search tree may contain an entire record of data, the definition of the tree depends on a given field whose values are distinct and may be ordered.

Definition: Suppose T is a binary tree. Then T is called a *binary search tree* if each node N of T has the following property:

The value of N is greater than every value in the left subtree of N and is less than every value in the right subtree of N.

It is not difficult to see that the above property guarantees that the inorder traversal of T will yield a sorted listing of the elements of T.

Remark: The above definition of a binary search tree assumes that all the node values are distinct. There is an analogous definition of a binary search tree T which admits duplicates, that is, in which each node N has the following properties:

(a) $N > M$ for every node M in a left subtree of N.
(b) $N \leq M$ for every node M in a right subtree of N.

When this definition is used, the operations discussed below must be modified accordingly.

(*a*) **Searching and Inserting in a Binary Search Tree**: The following is a search and insertion algorithm in a binary search tree *T*.

Algorithm 7.1 A binary search tree *T* and an ITEM of information is given. The algorithm finds the location of ITEM in *T*, or inserts ITEM as a new node in the tree.

Step 1. Compare ITEM with the root *N* of the tree.
  (*a*) If ITEM < *N*, proceed to the left child of *N*.
  (*b*) If ITEM > *N*, proceed to the right child of *N*.

Step 2. Repeat Step 1 until one of the following occurs:
  (*a*) We meet a node *N* such that ITEM = *N*. In this case the search is successful.
  (*b*) We meet an empty subtree, which indicates the search is unsuccessful. Insert ITEM in place of the empty subtree.

Step 3. Exit.

(*b*) **Deleting in a Binary Search Tree**: The following is an algorithm which deletes a given ITEM from a binary search tree *T*. It uses Algorithm 7.1 to find the location of ITEM in *T*.

Algorithm 7.2   A binary search tree T and an ITEM of information is given. $P(N)$ denotes the parent of a node N, and $S(N)$ denotes the inorder successor of N. The algorithm deletes ITEM from T.

Step 1.   Use Algorithm 7.1 to find the location of the node N which contains ITEM and keep track of the location of the parent node $P(N)$. (If ITEM is not in T, then STOP and Exit.)

Step 2.   Determine the number of children of N. There are three cases:

(a) N has no children. N is deleted from T by simply replacing the location of N in the parent node $P(N)$ by the NULL pointer.

(b) N has exactly one child M. N is deleted from T by replacing the location of N in the parent node $P(N)$ by the location of M. (This replaces N by M.)

(c) N has two children.

(i) Find the inorder successor $S(N)$ of N. (Then $S(N)$ has no left child.)

(ii) Delete $S(N)$ from T using (a) or (b).

(iii) Replace N by $S(N)$ in T.

Step 3.   Exit.

Remark: Observe that case (iii) in Step 2(c) is more complicated that the first two cases. The inorder successor $S(N)$ of N is found as follows. From the node N, move right to the right child of N, and then successively move left until meeting a node M with no left child. The node M is the inorder successor $S(N)$ of N.

Solved Problem 7.1 Suppose T is the binary tree stored in memory as in Figure 7-5. Draw the diagram of T.

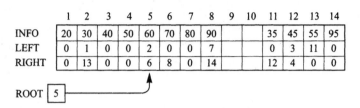

	1	2	3	4	5	6	7	8	9	10	11	12	13	14
INFO	20	30	40	50	60	70	80	90			35	45	55	95
LEFT	0	1	0	0	2	0	0	7			0	3	11	0
RIGHT	0	13	0	0	6	8	0	14			12	4	0	0

ROOT [5]

Figure 7-5

Solution. The tree T is drawn from its root R downward as follows:

(a) The root R is obtained from the value of the pointer ROOT. Note that ROOT = 5. Hence INFO[5] = 60 is the root R of T.

(b) The left child of R is obtained from the left pointer field of R. Note that LEFT[5] = 2. Hence INFO[2] = 30 is the left child of R.

(c) The right child of R is obtained from the right pointer field of R. Note that RIGHT[5] = 6. Hence INFO[6] = 70 is the right child of R.

We can now draw the top part of the tree as pictured in Figure 7-6(a). Repeating the above process with each new node, we finally obtain the required tree T in Figure 7-6(b).

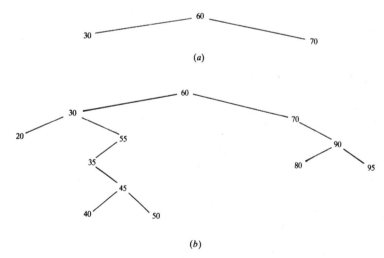

(a)

(b)

Figure 7-6

Chapter 8
BOOLEAN ALGEBRA

IN THIS CHAPTER:

✔ *Basic Definitions*
✔ *Duality*
✔ *Basic Theorems*
✔ *Logic Gates and Circuits*

Basic Definitions

Let B be a nonempty set with two binary operations + and *, a unary operation ', and two distinct elements 0 and 1. Then B is called a *Boolean algebra* if the following axioms hold where a, b, c are any elements in B:

[**B**$_1$]   Commutative laws:

$$a + b = b + a \qquad a * b = b * a$$

[**B**$_2$]   Distributive laws:

$$a + (b * c) = (a + b) * (a + c) \qquad a * (b + c) = (a * b) + (a * c)$$

[**B**$_3$]   Identity laws:

$$a + 0 = a \qquad a * 1 = a$$

[**B₄**]   Complement laws:

$$a + a' = 1 \qquad a * a' = 0$$

We will sometimes designate a Boolean algebra by $\langle B, +, *, ', 0, 1 \rangle$ when we want to emphasize its six parts. We say 0 is the *zero* element; 1 is the *unit* element, and a' is the *complement* of a. We will usually drop the symbol * and use juxtaposition instead. Then

$$a * (b + c) = (a * b) + (a * c)$$

is written

$$a(b + c) = ab + ac$$

which is the familiar distributive property from algebra. However,

$$a + (b * c) = (a + b) * (a + c)$$

becomes

$$a + bc = (a + b)(a + c),$$

which is certainly not a usual identity in algebra.

The operations $+$, $*$, and $'$ are called sum, product, and complement, respectively. We adopt the usual convention that, unless we are guided by parentheses, $'$ has precedence over $*$, and $*$ has precedence over $+$. For example,

$$a + b * c \text{ means } a + (b * c) \text{ and not } (a + b) * c$$

$$a * b' \text{ means } a * (b') \text{ and not } (a * b)'$$

Subalgebras; Isomorphic Boolean Algebras

Suppose C is a nonempty subset of a Boolean algebra B. We say C is a *subalgebra* of B if C itself is a Boolean algebra (with respect to the operations of B). We note that C is a subalgebra of B if and only if C is closed under the three operations of B, i.e., $+$, $*$, $'$.

Two Boolean algebras B and B' are said to be *isomorphic* if there is a one-to-one correspondence $f: B \rightarrow B'$ which preserves the three operations, i.e., such that

$$f(a+b) = f(a) + f(b)$$
$$f(a*b) = f(a) * f(b)$$
$$f(a') = f(a)'$$

for any elements, a, b in B.

Duality

The *dual* of any statement in a Boolean algebra B is the statement obtained by interchanging the operations $+$ and $*$, and interchanging their identity elements 0 and 1 in the original statement. For example, the dual of

$$(1 + a) * (b + 0) = b \quad \text{is} \quad (0 * a) + (b * 1) = b$$

Observe the symmetry in the axioms of a Boolean algebra B. That is, the dual of the set of axioms of B is the same as the original set of axioms. Accordingly, the important principle of duality holds in B. Namely,

Theorem 8.1 (Principle of Duality):

The dual of any theorem in a Boolean algebra is also a theorem.

In other words, if any statement is a consequence of the axioms of a Boolean algebra, then the dual is also a consequence of those axioms since the dual statement can be proven by using the dual of each step of the proof of the original statement.

Basic Theorems

Using the axioms $[\mathbf{B}_1]$ through $[\mathbf{B}_4]$, the following theorem can be proved.

Theorem 8.2 Let a, b, c be any elements in a Boolean algebra B.

(i) Idempotent laws:

$$a + a = a \quad a * a = a$$

(ii) Boundedness laws:

$$a + 1 = 1 \quad a * 0 = 0$$

(iii) Absorption laws:

$$a + (a * b) = a \quad a * (a + b) = a$$

(iv) Associative laws:

$$(a + b) + c = a + (b + c) \quad (a * b) * c = a * (b * c)$$

Theorem 8.2 and our axioms still do not contain all of the properties of sets. The next two theorems give us the remaining properties.

Theorem 8.3 Let a be any element of a Boolean algebra B.

(i) (Uniqueness of Complement)
If $a + x = 1$ and $a * x = 0$, then $x = a'$.

(ii) (Involution Law)
$(a')' = a$

(iii) $0' = 1; \quad 1' = 0$

Theorem 8.4 (DeMorgan's laws)

$$(a + b)' = a' * b' \quad (a * b)' = a' + b'$$

Logic Gates and Circuits

Logic circuits (also called *logic networks*) are structures that are built up from certain elementary circuits called *logic gates*. Each logic circuit may be viewed as a machine L which contains one or more input devices and exactly one output device. Each input device in L sends a signal, specifically a *bit* (binary digit),

$$0 \quad \text{or} \quad 1$$

to the circuit L, and L processes the set of bits to yield an output bit. Accordingly, an n-bit sequence may be assigned to each input device, and L processes the input sequences one bit at a time to produce an n-bit output sequence. First, we define the logic gates, and then we investigate the logic circuits.

Logic Gates

There are three basic logic gates, which are described below. We adopt the convention that the lines entering the gate symbol from the left are input lines and the single line on the right is the output line.

(a) **OR Gate**: Figure 8-1(a) shows an OR gate with inputs A and B and output $Y = A + B$ where "addition" is defined by the "truth table" in Figure 8-1(b). Thus the output $Y = 0$ only when inputs $A = 0$ and $B = 0$. Such an OR gate may have more than two inputs. Figure 8-1(c) shows an OR gate with four inputs, A, B, C, D, and output $Y = A + B + C + D$. The output $Y = 0$ if and only if all the inputs are 0.

Suppose, for instance, the input data for the OR gate in Figure 8-1(c) are the following 8-bit sequences:

$$A = 10000101, \quad B = 10100001, \quad C = 00100100, \quad D = 10010101$$

The OR gate only yields 0 when all input bits are 0. This occurs only in the 2nd, 5th, and 7th positions (reading from left to right). This the output is the sequence $Y = 10110101$.

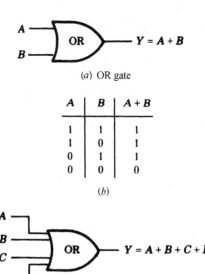

(a) OR gate

A	B	A + B
1	1	1
1	0	1
0	1	1
0	0	0

(b)

$Y = A + B + C + D$

(c)

Figure 8-1

(b) **AND Gate**: Figure 8-2(a) shows an AND gate with inputs A and B and output $Y = A \cdot B$ (or simply $Y = AB$) where "multiplication" is defined by the "truth table" in Figure 8-2(b). Thus the output $Y = 1$ only when inputs $A = 1$ and $B = 1$; otherwise $Y = 0$. Such an AND gate may have more than two inputs. Figure 8-2(c) shows an AND gate with four inputs, A, B, C, D, and output $Y = A \cdot B \cdot C \cdot D$. The output $Y = 1$ if and only if all the inputs are 1.

Suppose, for instance, the input data for the AND gate in Figure 8-2(c) are the following 8-bit sequences:

$A = 11100111, \quad B = 01111011, \quad C = 01110011, \quad D = 11101110$

The AND gate only yields 1 when all input bits are 1. This occurs only in the 2nd, 3rd, and 7th positions. This the output is the sequence $Y = 01100010$.

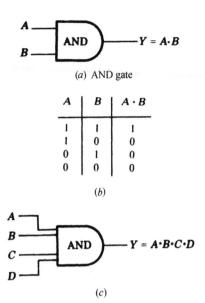

(a) AND gate

A	B	A · B
1	1	1
1	0	0
0	1	0
0	0	0

(b)

$Y = A \cdot B \cdot C \cdot D$

(c)

Figure 8-2

(c) **NOT Gate**: Figure 8-3(a) shows a NOT gate, also called an *inverter*, with input A and output $Y = A'$ where "inversion," denoted by the prime, is defined by the "truth table" in Figure 8-3(b). The value of the output $Y = A'$ is the opposite of the input A; that is, $A' = 1$ when $A = 0$ and $A' = 0$ when $A = 1$. We emphasize that a NOT gate can have only one input, whereas the OR and AND gates may have two or more inputs.

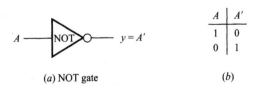

A	A'
1	0
0	1

(a) NOT gate   (b)

Figure 8-3

Suppose, for instance, a NOT gate is asked to process the following three sequences:

$$A_1 = 110001, \quad A_2 = 10001111, \quad A_3 = 101100111000$$

The NOT gate changes 0 to 1 and 1 to 0. Thus,

$$A_1' = 001110, \quad A_2' = 01110000, \quad A_3' = 010011000111$$

are the three corresponding outputs.

Logic Circuits

A logic circuit L is a well-formed structure whose elementary components are the above OR, AND, and NOT gates. Figure 8-4 is an example of a logic circuit with inputs A, B, C and output Y. A dot indicates a place where the input line splits so that its bit signal is sent in more than one direction. Working from left to right, we express Y in terms of the inputs A, B, C as follows. The output of the AND gate is $A \cdot B$, which is then negated to yield $(A \cdot B)'$. The output of the lower OR gate is $A' + C$, which is then negated to yield $(A' + C)'$. The output of the OR gate on the right, with inputs $(A \cdot B)'$ and $(A' + C)'$, gives us our desired representation, that is

$$Y = (A \cdot B)' + (A' + C)'$$

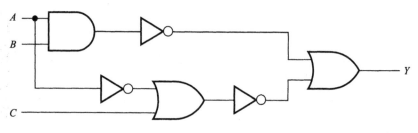

Figure 8-4

Logic Circuits as a Boolean Algebra

Observe that the truth tables for the OR, AND, and NOT gates are respectively identical to the truth tables for the propositions $p \lor q$ (disjunction, "p or q"), $p \land q$ (conjunction, "p and q"), and $\neg p$ (negation, "not p"). The only difference is that 1 and 0 are used instead of T and F. Thus, the logic circuits satisfy the same laws as do propositions and hence they form a Boolean algebra. We state this result formally.

Theorem 8.5 Logic circuits form a Boolean Algebra.

Accordingly, all terms used with Boolean algebras, such as complements, literals, fundamental products, miniterms, sum-of-products, and complete sum-of-products, may also be used with our logic circuits.

AND-OR Circuits

The logic circuit L which corresponds to a Boolean sum-of-products expression is called an AND-OR circuit. Such a circuit L has several inputs, where:

1. Some of the inputs or their complements are fed into each AND gate.
2. The outputs of all the AND gates are fed into a single OR gate.
3. The output of the OR gate is the output for the circuit L.

NAND and NOR Gates

There are two additional gates which are equivalent to combinations of the above basic gates.

(a) A NAND gate, pictured in Figure 8-5(a), is equivalent to an AND gate followed by a NOT gate.
(b) A NOR gate, pictured in Figure 8-5(b), is equivalent to an OR gate followed by a NOT gate.

The truth tables for these gates (using two inputs A and B) appear in Figure 8-5(c). The NAND and NOR gates can actually have two or more inputs just like the corresponding AND and OR gates. Furthermore, the

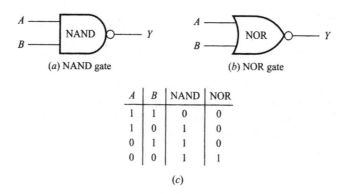

(a) NAND gate (b) NOR gate

A	B	NAND	NOR
1	1	0	0
1	0	1	0
0	1	1	0
0	0	1	1

(c)

Figure 8-5

output of a NAND gate is 0 if and only if all the inputs are 1, and the output of a NOR gate is 1 if and only if all the inputs are 0.

Observe that the only difference between the AND and NAND gates and between the OR and NOR gates is that the NAND and NOR gates are each followed by a circle. Some texts also use such a small circle to indicate a complement before a gate. For example, the Boolean expressions corresponding to the two logic circuits in Figure 8-6 are as follows:

(a) $Y = (A'B)'$, (b) $Y = (A' + B' + C)'$

(a) (b)

Figure 8-6

Solved Problem 8.1 Consider the Boolean algebra $\mathbf{D}_{210}$ (the set of divisors of 210).

(a) List its elements and draw its diagram.
(b) Find the set A of atoms.
(c) Find two subalgebras with eight elements.

Solution.

(*a*) The divisors of 210 are 1, 2, 3, 5, 6, 7, 10, 14, 15, 21, 30, 35, 42, 70, 105, and 210. The diagram of $\mathbf{D}_{210}$ appears in Figure 8-7.

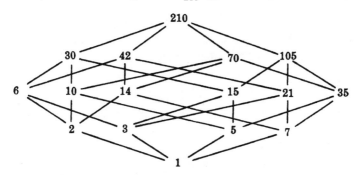

Figure 8-7

(*b*) $A = \{2, 3, 5, 7\}$, the set of prime divisors of 210.
(*c*) $B = \{1, 2, 3, 6, 35, 70, 105, 210\}$ and $C = \{1, 5, 6, 7, 30, 35, 42, 210\}$ are subalgebras of $\mathbf{D}_{210}$.

Solved Problem 8.2 Express the output *Y* as a Boolean expression in the inputs *A*, *B*, *C* for the logic circuit in Figure 8-8.

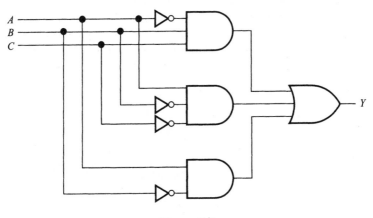

Figure 8-8

Solution. The output of the first AND gate is $A'BC$, of the second AND gate is $AB'C'$, and of the last AND gate is AB'. Thus,

$$Y = A'BC + AB'C' + AB'$$

Solved Problem 8.3 Express the output Y as a Boolean expression in the inputs A and B for the logic circuit in Figure 8-9.

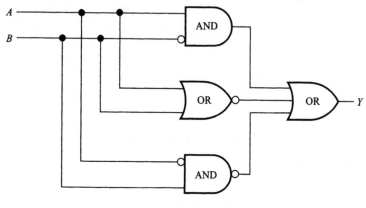

Figure 8-9

Solution. Here a small circle in the circuit means complement. Thus the output of the three gates on the left are AB', $(A + B)'$, and $(A'B)'$. Hence,

$$Y = AB' + (A + B)' + (A'B)'$$

Chapter 9

LANGUAGES, GRAMMARS, MACHINES

IN THIS CHAPTER:

- ✔ *Alphabet, Words, Free Semigroup*
- ✔ *Languages*
- ✔ *Grammars*
- ✔ *Finite State Machines*

Alphabet, Words, Free Semigroup

Consider a nonempty set A of symbols. A *word* or *string* w on the set A is a finite sequence of its elements. For example, the sequences

$$u = ababb \quad \text{and} \quad v = accbaaa$$

are words on $A = \{a, b, c\}$. When discussing words on A, we frequently call A the *alphabet*, and its elements are called *letters*. We will also abbreviate our notation and write a^2 for aa, a^3 for aaa, and so on. Thus, for the above words, $u = abab^2$ and $v = ac^2ba^3$.

The empty sequence of letters, denoted by λ, (Greek letter lambda) or ε (Greek letter epsilon), or 1, is also considered to be a word on A, called the *empty word*. The set of all words on A is denoted by A^*.

The *length* of a word u, written $|u|$ or $l(u)$, is the number of elements in its sequence of letters. For the above words u and v, we have $l(u) = 5$ and $\lambda(v) = 7$. Also, $l(\lambda) = 0$, where λ is the empty word.

Remark: Unless otherwise stated, the alphabet A will be finite, the symbols u, v, w will be reserved for words on A, and the elements of A will come from the letters a, b, c.

Concatenation

Consider two words u and v on alphabet A. The *concatenation* of u and v, written uv, is the word obtained by writing down the letters of u followed by the letters of v. For example, for the above words u and v, we have

$$uv = ababbaccbaaa = abab^2ac^2ba^3$$

As with letters, we define $u^2 = uu$, $u^3 = uuu$, and, in general, $u^{n+1} = uu^n$, where u is a word.

Clearly, for any words u, v, w, the words $(uv)w$ and $u(vw)$ are identical, they simply consist of the letters of u, v, w written down one after the other. Also, adjoining the empty word before or after a word u does not change the word u. In other words:

Theorem 9.1: The concatenation operation for words on an alphabet A is associative. The empty word λ is an identity element for the operation.

Subwords; Initial Segments

Consider any word $u = a_1a_2\ldots a_n$ on an alphabet A. Any sequence $w = a_ja_{j+1}\ldots a_k$ is called a *subword* of u. In particular, the subword $w = a_1a_2\ldots a_k$, beginning with the first letter of u, is called an *initial segment* of u. In other words, w is a subword of u if $u = v_1wv_2$, and w is an initial segment of u if $u = wv$. Observe that λ and u are both subwords of u since $u = \lambda u$.

Consider the word $u = abca$. The subwords and initial segments of u follow:

(1) Subwords: λ, a, b, c, ab, bc, ca, abc, bca $abca$
(2) Initial segments: λ, a, ab, abc, $abca$

Observe that the subword $w = a$ appears in two places in u. The word ac is not a subword of u even though all its letters belong to u.

Free Semigroup; Free Monoid

Let F denote the set of all nonempty words from an alphabet A with the operation of concatenation. As noted above, the operation is associative. F is called the *free semigroup over* A or the *free semigroup generated by* A . One can easily show that F satisfies the right and left cancelation laws. However, F is not commutative when A has more than one element. We will write F_A for the free semigroup over A when we want to specify the set A.

Now let $M = A^*$ be the set of words from A including the empty word λ. Since λ is an identity element for the operation of concatenation, M is called the *free monoid* over A.

Languages

A *language L over* an alphabet A is a collection of words on A. Recall that A^* denotes the set of all words on A. Thus a language L is simply a subset of A^*.

Operations on Languages

Suppose L and M are languages over an alphabet A. Then the "concatenation" of L and M, denoted by LM, is the language defined as follows:

$$LM = \{uv : u \in L, v \in M\}$$

That is LM denotes the set of all words which come from the concatenation of a word from L with a word from M. Clearly, the concatenation of languages is associative since the concatenation of words is associative.

Powers of a language L are defined as follows:

$$L^0 = \{\lambda\}, \quad L^1 = L, \quad L^2 = LL, \quad L^{m+1} = L^m L \text{ for } m > 1.$$

The unary operation $L*$ (read "L star") of a language L, called the *Kleene closure* of L, is defined as the infinite union

$$L^* = L^0 \cup L^1 \cup L^2 \cup \ldots = \bigcup_{k=0}^{\infty} L^k$$

Grammars

Figure 9-1 shows the grammatical construction of a specific sentence. Observe that there are:

  (1) various variables, e.g. ⟨sentence⟩, ⟨noun phrase⟩,…;
  (2) various terminal words, e.g., "The," "boy,"…;
  (3) a beginning variable ⟨sentence⟩; and
  (4) various substitutions or productions, e.g.,

⟨sentence⟩ → ⟨noun phrase⟩⟨verb phrase⟩
⟨object phrase⟩ → ⟨article⟩⟨noun⟩
⟨noun⟩ → apple

The final sentence only contains terminals, although both variables and terminals appear in its construction by the productions. The intuitive de-

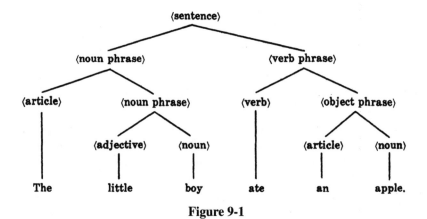

Figure 9-1

scription is given in order to motivate the following definition of a grammar and the language it generates.

A *phrase structure grammar* or, simply, a *grammar* G consists of four parts:

(1) A finite set (*vocabulary*) V.
(2) A subset T of V whose elements are called *terminals*; the elements of $N = V \setminus T$ are called *nonterminals* or *variables*.
(3) A nonterminal symbol S called the *start* symbol.
(4) A finite set P of productions. A production is an ordered pair (α, β) usually written $\alpha \rightarrow \beta$, where α and β are words in V. Each production in P must contain at least one nonterminal on its left side.

Such a grammar G is denoted by $G = G(V, T, S, P)$ when we want to indicate its four parts.

The following notation, unless otherwise stated or implied, will be used for our grammars. Terminals will be denoted by italic lowercase Latin letters, a, b, c, …, and nonterminals will be denoted by italic capital Latin letters, A, B, C, … with S as the start symbol.. Also, Greek letters, $\alpha, \beta, \ldots$, will denote words in V, that is, words in terminals and nonterminals. Furthermore, we will write

$$\alpha \rightarrow (\beta_1, \beta_2, \ldots, \beta_k) \text{ instead of } \alpha \rightarrow \beta_1, \alpha \rightarrow \beta_2 \ldots, \alpha \rightarrow \beta_k$$

Remark: Frequently, we will define a grammar G by only giving its productions, assuming implicitly that S is the start symbol and that the terminals and nonterminals of G are only those appearing in the productions.

Language $L(G)$ of a Grammar G

Suppose w and w' are words over the vocabulary set V of a grammar G. We write $w \Rightarrow w'$ if w' can be obtained from w by using one of the productions; that is, if there exists words u and v such that $w = u\alpha v$ and $w' = u\beta v$ and there is a production $\alpha \rightarrow \beta$. We write

$$w \Rightarrow\Rightarrow w' \text{ or } w^* \Rightarrow w'$$

if w' can be obtained from w using a finite number of productions.

Now let G be a grammar with terminal set T. The language of G, denoted by $L(G)$, consists of all words in T that can be obtained from the start symbol S by the above process; that is,

$$L(G) = \{w \in T^* : S \Rightarrow\Rightarrow w\}$$

Types of Grammars

Grammars are classified according to the kinds of production which are allowed. The following grammar classification is due to Noam Chomsky.

A Type 0 grammar has no restriction on its productions. Types 1, 2, and 3 are defined as follows:

(1) A grammar G is said to be of Type 1 if every production is of the form $\alpha \rightarrow \beta$ where $|\alpha| \leq |\beta|$ or of the form $\alpha \rightarrow \lambda$.
(2) A grammar G is said to be of Type 2 if every production is of the form $A \rightarrow \beta$, where the left side is a nonterminal.
(3) A grammar G is said to be of Type 3 if every production is of the form $A \rightarrow a$ or $A \rightarrow a\,B$, i.e., where the left side is a single nonterminal and the right side is a single terminal or a terminal followed by a nonterminal, or of the form $S \rightarrow \lambda$.

Observe that the grammars form a hierarchy; that is, every Type 3 grammar is a Type 2 grammar, every Type 2 grammar is a Type 1 grammar, and every Type 1 grammar is a Type 0 grammar.

Grammars are also classified in terms of context-sensitive, context-free, and regular as follows.

(a) A grammar G is said to be *context-sensitive* if the productions are of the form

$$\alpha A \alpha' \rightarrow \alpha \beta \alpha'$$

The name "context-sensitive" comes from the fact that we can replace the variable A by β in a word only when A lies between α and α'.

(b) A grammar G is said to be *context-free* if the productions are of the form

$$A \rightarrow \beta$$

The name "context-free" comes from the fact that we can now re-place the variable A by β regardless of where A appears.

(c) A grammar G is said to be *regular* if the productions are of the form

$$A \rightarrow a, \quad A \rightarrow aB, \quad S \rightarrow \lambda$$

Observe that a context-free grammar is the same as a Type 2 grammar, and a regular grammar is the same as a Type 3 grammar.

Solved Problem 9.1 Determine the type of grammar G which consists of the productions:

(a) $S \rightarrow aA, A \rightarrow aAB, B \rightarrow b, A \rightarrow a$
(b) $S \rightarrow aAB, AB \rightarrow bB, B \rightarrow b, A \rightarrow aB$
(c) $S \rightarrow aAB, AB \rightarrow a, A \rightarrow b, B \rightarrow AB$
(d) $S \rightarrow aB, B \rightarrow bA, B \rightarrow b, B \rightarrow a, A \rightarrow aB, A \rightarrow a$

Solution.

(a) Each production is of the form $A \rightarrow \alpha$, i.e., a variable on the left; hence G is a context-free or Type 2 grammar.
(b) The length of the left side of each production does not exceed the length of the right side; hence G is a Type 1 grammar.
(c) The production $AB \rightarrow a$ means G is a Type 0 grammar.
(d) G is a regular or Type 3 grammar since each production has the form $A \rightarrow a$ or $A \rightarrow aB$.

Finite State Machines

A *finite state machine* (or *complete sequential machine*) M consists of six parts:

(1) A finite set A of input symbols.
(2) A finite set S of "internal" states.
(3) A finite set Z of output symbols.
(4) An initial state s_0 in S.
(5) A next-state function f from $S \times A$ into S.
(6) An output function g from $S \times A$ into Z.

Such a machine M is denoted by

$$M = M(A, S, Z, s_0, f, g)$$

when we want to indicate its six parts.

Example. The following defines a finite state machine M with two input symbols, three internal states, and three output symbols:

(1) $A = \{a, b\}$.
(2) $S = \{s_0, s_1, s_2\}$.
(3) $Z = \{x, y, z\}$.
(4) Initial state s_0.
(5) Next-state function $f: S \times A \rightarrow S$ defined by

$$f(s_0, a) = s_1, \quad f(s_1, a) = s_2, \quad f(s_2, a) = s_0$$
$$f(s_o, b) = s_2, \quad f(s_1, b) = s_1, \quad f(s_2, b) = s_1$$

(6) Output function $g: S \times A \rightarrow Z$ defined by

$$g(s_0, a) = x, \quad g(s_1, a) = x, \quad g(s_2, a) = z$$
$$g(s_o, b) = y, \quad g(s_1, b) = z, \quad g(s_2, b) = y$$

Solved Problem 9.2 Consider the words

$$u = a^2 b a^3 b^2 \quad \text{and} \quad v = bab^2$$

Find: (a) uv;   (b) vu;   (c) v^2.

Solution. Write the letters of the first word followed by the letters of the second word:

(a) $uv = \left(a^2ba^3b^2\right)\left(bab^2\right) = a^2ba^3b^3ab^2$

(b) $vu = \left(bab^2\right)\left(a^2ba^3b^2\right) = bab^2a^2ba^3b^2$

(c) $v^2 = vv = \left(bab^2\right)\left(bab^2\right) = bab^3ab^2$

Solved Problem 9.3 What, if any, is the difference between the free semigroup on an alphabet A and the free monoid on A?

Solution. The free semigroup on A is the set of all nonempty words in A under the operation of concatenation; it does not include the empty word λ. On the other hand, the free monoid on A does include the empty word λ.

Index

117